# Planetary Ring Systems

Ellis D. Miner, Randii R. Wessen an
Jeffrey N. Cuzzi

# Planetary Ring Systems

Published in association with
**Praxis Publishing**
Chichester, UK

Dr. Ellis D. Miner
NASA Jet Propulsion Laboratory
(Retired)
Lake View Terrace
California
USA

Dr. Randii R. Wessen
NASA Jet Propulsion Laboratory (Caltech)
Pasadena
California
USA

Dr. Jeffrey N. Cuzzi
Research Scientist
Ames Research Center
NASA
Moffett Field
California
USA

---

SPRINGER–PRAXIS BOOKS IN SPACE EXPLORATION
SUBJECT *ADVISORY EDITOR*: John Mason, M.Sc., B.Sc., Ph.D.

---

ISBN 10: 0-387-34177-3 Springer Berlin Heidelberg New York

Springer is part of Springer-Science + Business Media (springer.com)

Bibliographic information published by Die Deutsche Bibliothek

Die Deutsche Bibliothek lists this publication in the Deutsche Nationalbibliografie; detailed bibliographic data are available from the Internet at http://dnb.ddb.de

Library of Congress Control Number: 2006933085


Printed in Germany



Cover design: Jim Wilkie
Project management: Originator Publishing Services, Gt Yarmouth, Norfolk, UK

Printed on acid-free paper

# Contents

# Authors' preface

All three of the authors of this book worked together on NASA's Voyager Project during its encounters with the four giant planets. Two of us (Miner and Cuzzi) continued that association during the preparation, launch, and early flight of the international Cassini–Huygens Mission to Saturn and Titan. The third (Wessen) is now Program Systems Engineer for Mission Systems Concepts at NASA's Jet Propulsion Laboratory. All three of us have enjoyed talking to general and professional audiences about the benefits and findings of the ongoing space program. While only one of us (Cuzzi) remains actively involved in ring studies (in fact serves as the Cassini Interdisciplinary Scientist for Rings), all of us continue to follow the successes and rich data return of the international space program, in particular that of the Cassini–Huygens Mission.

This book follows the general format and content of two prior books, *Uranus: The Planet, Rings and Satellites*, written as a part of the Wiley-Praxis Series in Astronomy and Astrophysics, and *Neptune: The Planet, Rings and Satellites*, a component of Springer-Praxis Books in Astronomy and Space Sciences. In fact, some of the material in those two books, to the extent that it is still current and relevant to the contents of this book, has been used in the discussion of the Uranus and Neptune ring systems. It is our belief that the inspiring story of the continuing discoveries associated with spacecraft and telescopic exploration of the solar system cannot be told too often. Although much of the planning, data collection, and data analysis lies in the hands of professionals, the harvest belongs to the world, and it is our sincere desire to make these findings as accessible as possible to those who are not professional ring scientists. Most of the research, and indeed all the elements of the space program relevant to the contents of this book, is publicly funded, primarily by the established space agencies of the United States and other nations. We are grateful that so many nations have chosen to support ongoing space research, ground-based telescopic observations, and theoretical studies.

A number of professional texts published by the University of Arizona Press contain articles about the ring systems. Both *Saturn*, edited by Tom Gehrels and Mildred S. Matthews, and *Planetary Rings*, edited by Richard Greenberg and André Brahic, were published in 1984. They have been referenced frequently in this book, particularly in discussions of pre-Galileo knowledge of Jupiter's ring system and pre-Cassini knowledge of Saturn's ring system. Both were published before the Voyager encounters of Uranus and Neptune. *Uranus*, edited by Jay T. Bergstralh, Ellis D. Miner, and Mildred S. Matthews (published in 1991), and *Neptune and Triton*, edited by Dale P. Cruikshank (published in 1995), have been invaluable sources for discussions of the ring systems of those two planets. The importance of spacecraft observation is so great that these books remain a good introduction to the subject decades after their publication.

The advent of the Hubble Space Telescope and of adaptive optics systems in large ground-based telescopes, and the development of radar and passive radio systems at large radio telescopes (especially Arecibo in Puerto Rico) have also provided much new data on the rings that supplement and update the data from Voyagers 1 and 2. Most of those results are available in open scientific literature. Advances in theoretical studies of ring systems also benefit from the better understanding of ring system characteristics that is emerging from the observations. It is apparent that the ring systems, particularly (but not limited to) that of Saturn, are far more complex than anyone had reason to believe a few short decades ago.

It was in this environment that the authors were approached about producing a text describing the known planetary ring systems and aimed at general audiences. One of us (Miner) was at first reluctant to do so, in part because of approaching retirement, but he took only a little additional persuasion. It has been challenging for each of us to do the research and writing necessary to bring this project to fruition in the midst of other competing responsibilities. For those of you unfamiliar with the world of scientific books, such projects provide little financial remuneration; this book is presented to you, fellow citizens of Earth, as a labor of love.

*Ellis D. Miner* (NASA-JPL retired)
*Randii R. Wessen* (NASA-JPL)
*Jeffrey N. Cuzzi* (NASA-ARC)

*California, July 2006*

# Acknowledgments

The authors gratefully acknowledge the National Aeronautics and Space Administration and two of its centers, Caltech's Jet Propulsion Laboratory in Pasadena and Ames Research Center in Silicon Valley, for most of the figures contained in this book.

*Dedicated to our colleagues,*
*past and present,*
*whose planetary ring studies*
*made this book possible*

# List of figures

# List of tables

# List of abbreviations and acronyms

| | |
|---|---|
| AU | Astronomical Unit |
| CDA | Cassini Dust Analyzer |
| CIRS | Thermal InfraRed telescope |
| DSN | Deep Space Network |
| DTR | Digital Tape Recorder |
| ESA | European Space Agency |
| GEM | Galileo Europa Mission |
| GMM | Galileo Millennium Mission |
| GMME | Galileo Millennium Mission Extension |
| HEF | High-Efficiency Antenna |
| HGA | High-Gain Antenna |
| HST | Hubble Space Telescope |
| IMC | Image Motion Compensation |
| IPP | Imaging PhotoPolarimeter |
| IRIS | InfraRed Interferometer Spectrometer and radiometer |
| ISS | International Space Station; Imaging Science Subsystem |
| JPL | Jet Propulsion Laboratory |
| JRINGS | Single-ring search observation |
| JWST | James Webb Space Telescope |
| KAO | NASA's Gerard P. Kuiper Airborne Observatory |
| LGA | Low Gain Antenna |
| MIMC | Maneuverless Image Motion Compensation |
| NASA | National Aeronautics and Space Agency |
| NIMC | Nodding Image Motion Compensation |
| NIRC | Near InfraRed Camera |
| NRAO | National Radio Astronomy Observatory |
| ORS | Optical Remote Sensing |
| PPS | PhotoPolarimeter Subsystem |

| | |
|---|---|
| PST | Pacific Standard Time |
| RPWS | Radio and Plasma Wave Spectrometer |
| RSS | Radio Science Subsystem |
| SOI | Saturn Orbit Insertion |
| TNT | TriNitroToluene |
| UV | UltraViolet |
| UVIS | UltraViolet Imaging Spectrograph |
| UVS | UltraViolet Spectrometer |
| VIMS | Visible and Infrared Mapping Spectrometer |

# 1

# The scientific significance of planetary ring systems

## 1.1 "SMAISMRMILMEPOETALEUMIBUNENUGTTAUIRAS"

Italian astronomer Galileo announced with this anagram sent to his patron astronomer Johannes Kepler on July 30, 1610, that (unscrambling his anagram), "ALTISSIMUM PLANETAM TERGEMINUM OBSERVAVI." Roughly interpreted out of the Latin, Galileo was telling his colleague, "I have observed that the most distant of planets has a triple form." (Note that U and V are interchangeable in Latin.) Of course, Kepler did not understand what Galileo was announcing until Galileo himself later unscrambled the anagram. This was a practice used by early astronomers to lay claim to a new discovery without divulging its nature until the results were ready for publication. Galileo had been the first to observe the rings of Saturn, although their basic particulate nature (Figure 1.1) was not recognized until the middle of the 19th Century.

Four centuries have gone by since Galileo's startling discovery. The existence of Saturn's extensive ring system (and lesser ring systems around Jupiter, Uranus, and Neptune) is no longer debated. Nevertheless, as more and more details of these ring systems surface, the observations reveal complexities never before imagined, and many of the features within the rings have not yet been adequately explained. In that sense, this book may be somewhat premature. It is our intent to provide possible explanations for many of the observations as well as to describe phenomena that remain unexplained. Scientific advances often come from the ranks of those who may not be experts in the field, but whose interest leads them to contemplate such mysteries. Perhaps among the readers of this book there will be some whose new insights about ring dynamics may one day become the accepted explanations for those phenomena.

In a sense, the rings first viewed by Galileo in 1610 and later observed in greater and greater detail by Earth-based telescopes and sophisticated spacecraft have presented us an anagram far more difficult to interpret than Galileo's Latin one. Of

**Figure 1.1.** This series of images from the Hubble Space Telescope was obtained over the years 1996 through 2000, showing the ever-changing appearance of Saturn's ring system. The view at the lower left (1996) is approximately the same ring opening as when Galileo first observed Saturn in 1610. By squinting the eyes, it is not difficult to see why the fuzzy view through his small telescope might have led him to think there were two smaller bodies flanking Saturn. (PIA03156)

course, we have the handicap that Nature seldom intervenes to provide us clear and concise interpretations. Instead, we are left to interpret the observations through years of intensive study and contemplation. Of one thing we can be certain: the giant planet ring systems do not violate the laws of physics. But in the process of ascertaining that fact, we may come to newer and more complete understanding of those immutable laws. We welcome you to share in that adventure.

## 1.2 FIRST IMPRESSIONS

Almost everyone who has viewed Saturn and its rings through a moderate-sized telescope can remember that first experience vividly. Many long years ago, the wife of one of the authors of this book was expecting their first child. About two weeks before the due date, we had occasion to view Saturn through a 24-inch Cassegrain

**Figure 1.2.** This view shows BYU's Eyring Science Center, capped by the observatory which housed a 24-inch reflecting telescope in the early 1960's. It was through this telescope that one of the authors and his wife first viewed the rings of Saturn. (Image courtesy of Brigham Young University).

telescope atop the Eyring Science Center on the campus of Brigham Young University (see Figure 1.2). There were several other young couples with us on that occasion, and, like us, one of the other couples was expecting a child in the near future. The sky clarity and stillness were ideal, and Saturn made a huge impression on each of us. The night was June 28, 1962, and before the evening hours of June 29 both expectant mothers had given birth to healthy new babies. It is entirely possible that our viewing of Saturn that evening had no connection with the births of those two babies; it is also likely that the climbing of a steep flight of stairs that marked the entrance to the observatory was more directly responsible for the timing of these blessed events than was Saturn. Whichever (if either) hastened the births of the two babies, Saturn has nevertheless held a special place in our family's hearts and minds since that night.

Because of its spectacular appearance through even a moderate-sized telescope, primarily due to the presence of its bright ring system, Saturn has often been called the "Jewel of the Solar System." For more than three and a half centuries, Saturn's ring was the only planetary ring system known to humanity. It is still the only planetary ring system easily viewable from Earth-based telescopes. The rings of Uranus were discovered in 1977 from Earth-based telescopes as each ring momentarily blocked the light of a background star in front of which Uranus was passing. Similar star "occultation" measurements of Neptune from Earth left the nature of its rings uncertain at best, sometimes returning positive results, but more often showing no evidence of a Neptune ring system. The discovery of the Jupiter ring system in 1979, the first images of the Uranus ring system, and the verification of a Neptune ring system in 1989 were left for NASA's hardy interplanetary spacecraft, Voyagers 1 and 2. The Galileo mission provided much additional data on the precise nature and origin of Jupiter rings between 1995 and 2003. In a similar fashion, data on the Saturn ring system from the Cassini Orbiter between 2004 and 2008 outshine by far any prior data from Earth or from interplanetary spacecraft. An entire chapter of this book (Chapter 10) will be devoted to a preliminary analysis of Saturn ring system data from the Cassini mission.

## 1.3 MAJOR CHARACTERISTICS SHARED BY THE KNOWN RING SYSTEMS

At first blink, one might suppose that all the known planetary ring systems are pretty much alike. For example, essentially all the ring particles at each of the four giant planets orbit their respective planet very close to that planet's equator. All are composed primarily of small particles, with a major fraction of the ring particles having typical sizes of a centimeter or less. Most of the rings are very thin compared with their radial extent, two major exceptions being Saturn's outermost ring, called the E ring, and Jupiter's innermost ring, called the Halo ring (see Figure 1.3). Both the E ring and the Halo ring are composed of tiny particles, most with diameters of about a micrometer, or about the size of the particles in cigarette smoke. Particles with such small sizes (and correspondingly small mass) may be subject to other forces which, over the long term, may substantially alter their orbits around the central planet. The possible nature of those forces will be discussed in later chapters.

All of the known ring systems are relatively young compared with the ages of their respective planets. The estimated ages of the four giant planets is about $4.5 \times 10^9$ years, close to the estimated ages of both the Earth and the Sun. A number of factors seem to indicate that dynamic processes within the four ring systems limit their lifetimes to no more than 1% (and perhaps closer to 0.1%) of the ages of the planets. Rings composed of dust-sized particles may have lifetimes that are much shorter,

**Figure 1.3.** In the long-exposure upper image, the vertical extent of Jupiter's Halo ring can be seen to extend well above and below the equatorial plane; in the lower image, the Main ring lies well within a degree of the equatorial plane. (PIA01622)

perhaps as short as a human lifetime. We either have the fortune to be living at precisely the time when rings exist for each of the giant planets, an unlikely scenario, or each of the rings is replenished by ongoing processes. In later chapters, we will try to outline some of these replenishment processes, which vary somewhat from planet to planet.

Each of the ring systems seems to exhibit a structure which has the appearance of a series of concentric ringlets, with easily observable radial structure and little or no observable azimuthal variations. This radial sorting of ring particles is primarily the result of gravitational interaction with nearby planetary satellites, although there are a variety of ways this interaction affects the structure. Sometimes it causes sharp inner or outer boundaries of a ring. At other times gaps in an otherwise (radially) continuous ring are created. Gravitational forces from nearby satellites can even result in effectively "corrugating" the rings (bending waves) or causing tightly wound spiral variations in the ring particle population (density waves). We will attempt to explain these effects (at least those that are well understood) in terms that do not require the reader to have an extensive background in celestial dynamics or a keen understanding of higher mathematics. All in all, the creation and molding of planetary ring systems is both complex and fascinating.

## 1.4 MAJOR DIFFERENCES BETWEEN THE KNOWN RING SYSTEMS

While the similarities between the ring systems are primarily those that provide a description of planetary ring systems in general, it is the differences that give each ring system its own unique personality and appearance. It is also these individually unique personalities that provide the motivation for this book. Without that diversity, a treatise on planetary ring systems might require little more than a 15-page illustrated magazine article. In this introductory chapter, only a few of the major differences between the ring systems are discussed. In later chapters the individual ring systems will be discussed in detail.

Composition differences between the ring systems are substantial. Jupiter's ring system is primarily rocky (silicate) material, tiny fragments of several of the small inner moons of Jupiter. The Jupiter ring system appears to be almost completely devoid of icy material, either water ice ($H_2O$) or methane ice ($CH_4$). The Saturn ring system, on the other hand, is dominated by water ice, although other constituents, especially rocky materials, are also relatively abundant. The composition of the Uranus and Neptune rings is inferred more on the basis of theoretical modeling than on actual spectral measurements. The Voyager 2 spacecraft, the sole planetary probe to visit these two distant giant planets, did not carry a mass spectrometer, and the narrow rings of Uranus and Neptune were difficult targets for the relatively broad field of view of Voyager 2's infrared spectrometer. However, it is tempting to conclude, because of their darkened surfaces, that the ring particles in these two ring systems are either coated with carbon or composed largely of carbon. The theoretical source of such carbon is methane, which is more abundant in the atmospheres of Uranus and Neptune than in either Jupiter's or Saturn's atmospheres. Methane ice may also be abundant in the icy moons of Uranus and Neptune. Processes which might cause the methane gas or ice to separate into its constituent elements—carbon and hydrogen—have also been proposed. (These processes will be discussed in a later chapter.) Once the carbon and hydrogen are separated, the hydrogen, with its low mass, could more easily escape into interplanetary space, thereby leaving behind it an excess of free carbon to darken the surfaces of the ring particles.

The ring systems also differ markedly in their physical appearance. Jupiter's ring is confined relatively close to the planet. Because it is primarily composed of dispersed tiny dust particles, the Jupiter ring is almost transparent. In that respect, it is much like dust on an automobile windshield, barely visible except when looking through it toward the Sun. Jupiter's "Main ring" and its "Gossamer ring" (the nomenclature will be explained in Chapter 4) are flat and very near the plane of Jupiter's equator; the "Halo" is between the other rings and Jupiter itself, but is spread north and south of Jupiter's equator into an enormous donut shape.

Saturn's ring system is the most extensive in a radial direction, reaching nearly to the orbit of its moon Titan, a distance nearly twenty times the radius of the planet. Most of Saturn's rings are thin and near the equator of the planet, but Saturn's outermost ring, its E ring, reaches a vertical thickness of thousands of kilometers in

**Figure 1.4.** Voyager 2 captured this image of two of Neptune's narrow rings in 1989. The outer, or Adams, ring is seen to have several brighter ring arcs along a portion of its circumference. The overexposed crescent of Neptune is at the lower left. (PIA02207)

places. There are few empty gaps in Saturn's ring system, except where there are satellites that basically sweep up ring particles near their orbits.

The rings of Uranus and Neptune, by contrast, are mainly very narrow structures with little in the way of ring matter (other than tiny dust particles) between them. The innermost ring of Uranus is broad, without distinct edges, and difficult to see in Voyager or Earth-based images. Recent advances in instrumentation and image processing have now permitted the viewing of all of the Uranus rings by the Hubble Space Telescope and the Keck Telescope equipped with adaptive optics, including two new rings exterior to those imaged by Voyager 2, but still interior to the orbit of Miranda. Most of Uranus's rings are non-circular, and they are generally somewhat wider at the most distant parts of their orbits than at the closer distances. This is especially noticeable in the outermost of the rings observed by Voyager 2 (the Epsilon ring). The outermost (Adams) ring of Neptune (see Figure 1.4) sports five brighter ring arcs (Courage, Liberté, Egalité 1, Egalité 2, and Fraternité), all confined within about 10% of the ring circumference. One faint ring seems to share the orbit of one of Neptune's satellites (Galatea) and is sometimes called the Galatea ring. Another, known as the Lassell ring, seems to span the approximately 4,000 km between the narrow Arago and Le Verrier rings. Still another, the Galle ring, has a narrow core but diffuses outward and inward about 1,000 km, with no distinct outer or inner edge.

As is apparent from the above, it is patently untrue that "A ring is a ring is a ring!"—hence the need for a book like this one that provides a partial roadmap to the planetary ring systems of which we are presently aware. As new planetary ring systems are discovered and new characteristics of previously known planetary ring systems are disclosed, it will very likely be necessary to expand the contents of this book. In that sense, planetary ring systems are much like their larger cousins, the satellites circling the planets of our solar system: the closer we examine them, the more complex and varied they become. However, in many ways it is that very complexity that may eventually provide the key to a relatively complete understanding of their origins, their present characteristics, and their eventual fates.

## 1.5 WHAT CAUSES RING SYSTEMS?

Between 1610, when Galileo first glimpsed Saturn's ring, and 1977, when the ring of Uranus was discovered, the solar system, for that matter the universe, contained only a single example of a planetary ring system—namely, the ring of Saturn. Astronomers speculated on which of two mechanisms gave birth to that ring: (1) Was it the remnant material from Saturn's formative years that was prevented from coalescing into individual satellites? or (2) Was the ring material the debris from one or more former satellites that somehow wandered too close to Saturn and were torn apart by gravitational tides or were shattered by a collision with an interplanetary interloper? It is now abundantly clear that the first of these choices is no longer tenable, primarily because ring scientists have come to understand that the present rings cannot have existed for more than a small fraction of the age of the solar system.

Does that then once and for all answer the question about the origins of Saturn's rings and, by analogy, the origins of planetary ring systems in general? Unfortunately (or perhaps fortunately, depending on one's viewpoint), continued observations of planetary ring systems have suggested that a number of different mechanisms help to generate rings and define their shape. As is often the case in scientific investigations, more detailed examination of ring characteristics and phenomena revealed a wholly unsuspected complexity and diversity associated with planetary rings and ring systems.

Let's examine briefly the second mechanism proposed by astronomers: that the rings of Saturn formed as one or more satellites wandered too close to Saturn. Most individuals are familiar with the concept of tides, especially Earth's tides, which are caused primarily by the gravitational pull of Earth's only natural satellite, the Moon. What may be less familiar is that the Earth also exerts tidal forces on the Moon. The Moon has no liquid on its surface, but the solid surface is deformed as a result of these tidal forces. The Moon keeps its same face toward the Earth, so the deformation results in a slightly non-circular equator for the Moon, whose equatorial diameter in the line of sight from Earth is more than 600 meters larger than its equatorial diameter along its orbit. That equatorial deformation is only about 0.002% of the equatorial diameter.

French mathematician Édouard Albert Roche (1820–1883) recognized that, for a satellite which is closer than a distance which has become known as the "Roche limit", such tidal forces can become stronger than the forces which hold the satellite together. In such circumstances the satellite will be torn asunder. For an icy satellite circling Saturn, the Roche limit is at a distance from Saturn's center of about 2.4 times the radius of the planet. All of Saturn's main rings (D, C, B, A, and F) are within this distance, so it seems likely that they are primarily the product of tidal breakup of former moons of Saturn. The approximate distance (from the center of the respective planet) of the Roche limit for Earth and the four giant planets is given in Table 1.1. The equatorial radius of each planet is also shown. The main rings of Saturn and Uranus and all the rings of Jupiter and Neptune are within their respective Roche limits. Earth's Moon is at a mean distance from Earth's center of 384,400 km, more than twenty times Earth's Roche limit.

**Table 1.1.** Roche limit (in kilometers and miles) for Earth and the four giant planets.

| *Planet* | *Radius* (km) | *Roche limit* (km)* | *Roche limit* (mi)* |
|---|---|---|---|
| Earth | 6,378 | 18,470 | 11,470 |
| Jupiter | 71,492 | 175,000 | 108,000 |
| Saturn | 60,268 | 147,000 | 92,000 |
| Uranus | 25,559 | 62,000 | 39,000 |
| Neptune | 24,764 | 59,000 | 37,000 |

* Roche limit distances are referenced to the center of the respective planet.

The extended G and E rings of Saturn lie exterior to the main rings and also exterior to the Saturn Roche limit. They are therefore unlikely to have been formed by the breakup of natural satellites or gravitationally captured asteroids. The origin of the G ring remains unexplained, but the origin of the E ring has been attributed to icy eruptions from Saturn's satellite Enceladus. More will be said about that in Chapter 10, which covers early findings on the ring system from the Cassini Orbiter.

Once a large quantity of ring particles has been created, mutual collisions between the particles cause them to spread in all directions from the point of breakup. Polar flattening of the planet and the gravitational influence of satellites, nearly all of which orbit the planet near its equatorial plane, tend to flatten the distribution of the ring particles over time. Given enough time, the azimuthal distribution of ring particles becomes almost uniform unless the gravitational interactions with nearby satellites prevent such a uniform distribution. Radial spreading is also relatively efficient, but such spreading is inhibited or modified by two mechanisms. Those particles that spread inward toward the planet may eventually fall into the atmosphere of the planet and are lost from the ring system. Both outward and inward movement of particles can also be inhibited by the gravitational influence of nearby satellites. These two effects will be discussed in more detail in later chapters.

Jupiter's Main and Gossamer rings are unlikely to be the result of either satellite breakup or eruptions from satellite surfaces. In some ways, Jupiter's enormous gravity acts like a magnet for large numbers of meteoroids in the solar system. These meteoroids occasionally strike the surfaces of small satellites and launch surface dust and debris into orbit around Jupiter. The Gossamer ring actually has characteristics that enable scientists to determine relatively unambiguously the source satellite for the ring material. More will be said about this in Chapters 4 and 6.

Jupiter's Halo ring and tiny particles that form radial spokes in Saturn's B ring are affected by still another mechanism. If the particles are tiny enough and can also become electrically charged, either by the action of sunlight or by other mechanisms, such particles can be deflected from normal near-equatorial orbits around Jupiter or Saturn by the respective planet's magnetic field, which tends to drag them along at the angular speed of rotation of the planet itself. For Jupiter's Halo ring, the result is a donut-shaped torus of ring particles interior to Jupiter's main ring. For Saturn's B-ring spokes, the result is ghostly structures that seem to defy the laws of orbital mechanics.

It is likely that a variety of other mechanisms create and shape planetary rings, and it is partially the pursuit of ring studies to discover and explain these effects and to use them as predictors of ring system evolution.

## 1.6 WHAT RINGS TELL US ABOUT THE PLANET THEY CIRCLE

Knowledge of some characteristics of the central planet can be gleaned from or enhanced by careful measurements of planetary ring systems. Such gleaning or knowledge enhancement requires a clear understanding of the precise nature of the interactions between the planet and its ring system. Furthermore, there must exist a means of determining the relevant ring system characteristics with sufficient precision to enable the extraction of the desired central planet information. Let's look at a couple of examples.

Early ring observers, including Christiaan Huygens (1629–1695), who correctly deduced on the basis of long-term observations that the Saturn ring circled the planet, nowhere touching it, incorrectly assumed first that the ring was a solid disk of material and then later that there were a number of solid concentric rings. Both theoretical considerations (by James Clerk Maxwell in 1859) [1] and observational data (by James Keeler in 1895) [2] eventually convinced scientists that the rings must consist of an innumerable array of discrete ring particles, which, like the tiny satellites they were, orbited according to the laws of motion described by Johannes Kepler (1571–1630). According to those laws, independent bodies orbiting closer to the planet complete an orbit of the planet in less time than those further from the planet. In addition, close orbiters must also move at higher velocities than their more distant siblings. The rate of motion is more or less independent of the size and mass of the individual rings particles, but, as implied above, the rate is highly dependent on the radial distance of each particle from the center of the planet. Additionally, that speed is dependent on the mass of the planet whose gravity holds the ring particles in their orbits. Theoretically, then, it should be possible to deduce the mass of the planet from measurements of the distance and orbital periods of a large number of ring particles.

In practice, however, it is very difficult to measure the orbital periods of individual ring particles, primarily because of their small size and the impracticality of uniquely identifying individual ring particles at all, let alone uniquely re-identifying the same particles on later circuits of the planet. Mass determination for the central planet is far more easily determined from the orbital motions of the larger natural satellites, or, still better, from radio tracking of properly equipped robotic spacecraft during swingbys or orbits of the respective planet.

Although ring observations contribute little toward improvement of the precision of mass determinations of the planet, they can, under certain circumstances, help us to better understand the distribution of mass within the central planet. At the same time, they can help define the planet's equatorial plane. The orbit of a ring that is eccentric (non-circular) will slowly precess (turn) at a rate that depends directly on the degree of flattening (oblateness) of the planet. Hence, a measurement of the precession rate of an eccentric ring orbit can lead to a determination of the planet's oblateness. Note, however, that the oblateness determined in this manner is not necessarily the same as the amount of flattening seen in images of the planet (i.e., the *optical* oblateness), but instead represents the degree to which mass has been shifted from the polar regions of the planet toward its equator. That quantity is known as the planet's *dynamic* oblateness and is generally designated as $J_2$. A planet without polar flattening would have $J_2 = 0$. With a $J_2$ of 0.0163, Saturn is the most oblate planet in the solar system. The dynamic oblateness of Jupiter, Uranus, and Neptune are 0.0147, 0.0035, and 0.0034, respectively [3].

All four known planetary ring systems lie close to the equatorial planes of their respective planets, the largest departures of ring planes from planetary equatorial planes being less than a tenth of a degree—that is, less than 1/900th of a right angle. In that respect, rings provide a reasonably accurate visual marker for the planet's equator. But if the orientation of an inclined ring orbit can be tracked through a large part of a 360° precession, the actual orientation of the plane of the equator can often be determined to a small fraction of the ring's inclination—that is, to less than a hundredth of a degree.

Tiny dust particles in the rings also interact with the planetary magnetic field. Two specific examples of such interactions are radial spokes in Saturn's B ring and Jupiter's Halo ring. The B-ring spokes rotate, at least for a time, at the same rate as Saturn's magnetic field and therefore provide corroborating information on the rotation rate of Saturn's interior. The Halo ring has inner and outer boundaries near what are called Lorentz resonances, where electrically charged dust particles orbit Jupiter in periods that are rational multiples (like 3:2) of the rotational period of Jupiter's magnetic field (and the interior of the planet). At the radial distances of these resonances, the ring particles experience strong forces tending to remove them from the ring altogether, thus creating relatively sharp edges to these dust rings [4]. Studies of these phenomena can help determine the rotation rate of the planet's magnetic field or serve as a verification of the rotation rate as determined from radio emissions.

## 1.7 WHAT RINGS TELL US ABOUT THE NEARBY MOONS

Spiral bending waves (corrugations) and density waves (alternate crowding and separating of ring particles in a radial direction) are due to gravitational interactions with nearby satellites, especially in the rings of Saturn and Uranus (Figure 1.5). Sharp edges to ring boundaries or gaps within otherwise continuous rings are generally also indications of ring particle interactions with natural satellites. Still other effects—like wakes along an otherwise sharp edge or condensations within a ring, or chaotic appearance at the outer or inner boundaries of a ring—are due to interactions with planetary satellites.

It is therefore natural, when such ring features are observed, to try to associate the features with known satellites. Occasionally such associations with known satellites are not possible, or the ring features are of such a nature that they are more likely to be associated with nearby, as yet unseen, satellites. In such circumstances, the ring features often lead to predictions about the precise location of a potential new satellite, and additional observations (or poring over existing images) can result in the discovery of the proposed satellite. The former procedure led to the discovery of a small satellite (Daphnis) orbiting within the Keeler gap near the outer edge of Saturn's A ring. The latter is exemplified by Pan, discovered nearly eight years after the Voyager swingbys of Saturn from Voyager images of the Encke gap in Saturn's A ring [5].

The nature and magnitude of the gravitational interactions between ring particles and the satellites responsible for the individual features can also lead to a rough estimate of the mass of the perturbing satellite. When these are combined with images of sufficient resolution to permit size and shape estimates for the relevant satellite, rough density estimates are possible. Those density estimates can further lead to composition estimates, relatively easily distinguishing between solid rocky satellites

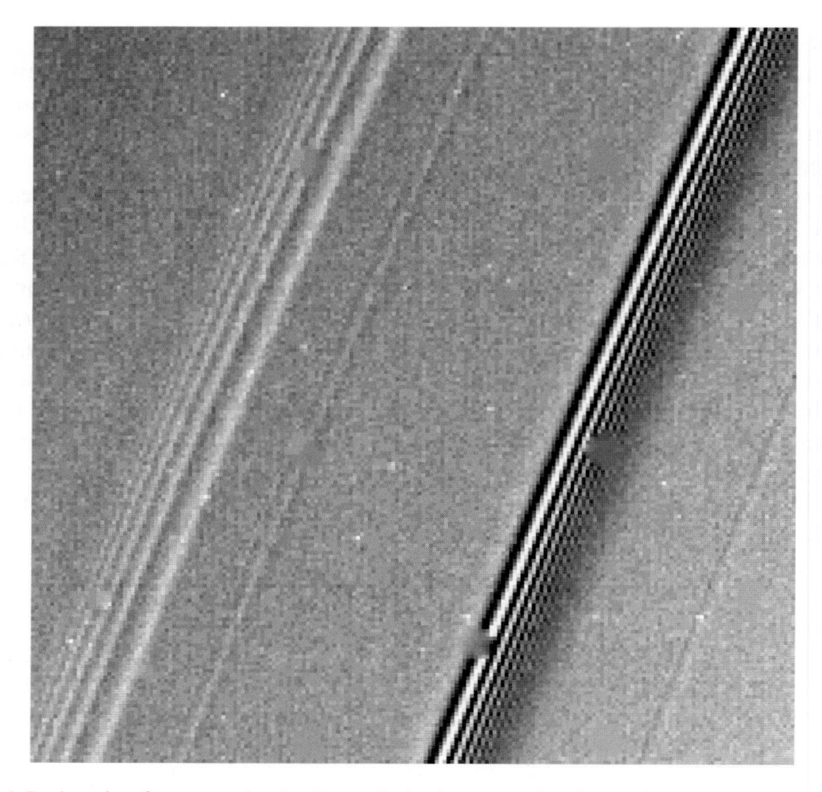

**Figure 1.5.** A pair of wave trains in Saturn's A ring near the distance at which ring particles circle the planet five times for every three orbits of the moon Mimas. At the left is a spiral density wave. At the right is a bending wave, where the ridges cast shadows in the adjacent troughs. The distance between the two wave trains is about 400 km. (Voyager FDS 43993.50)

(with densities greater than 2 grams per cubic cm, abbreviated $\mathrm{g\,cm^{-3}}$), solid icy satellites (densities between about 1 and $1.5\,\mathrm{g\,cm^{-3}}$), or loose "rubble piles" of icy material (densities significantly less than $1\,\mathrm{g\,cm^{-3}}$).

Finally, radial ring density variations and, to some extent, vertical ring structure, has led ring scientists to speculate that the source of nearby ring material is from a particular satellite. Saturn's expansive E ring was brightest (densest) near the orbit of Enceladus, and one of the goals of the Cassini Mission was to determine if evidence can be found to substantiate geologically active processes on Enceladus that might generate E-ring particles. Similarly, Jupiter's Gossamer ring has a vertical structure that has led scientists to speculate that it is being fed by particles being blasted (most likely by meteoroid bombardment) from the surfaces of Thebe and Amalthea, each of whose orbits is slightly inclined to Jupiter's equatorial plane. The Gossamer ring particles thus generated seem to migrate only toward Jupiter's atmosphere and not outward. That behavior and possible explanations for it will also be discussed in later chapters.

## 1.8 RINGS AS A NATURAL LABORATORY FOR MANY-BODY INTERACTIONS

One of the tougher problems in theoretical dynamics is the precise mathematical formulation of the interaction of a very large array of individual particles. Planetary ring systems are but one variation of this problem. Others, in order of increasing sizes, are the tens of thousands of asteroids in the solar system's asteroid belt, the Kuiper Belt of icy bodies in the outer solar system, the spherical array of comets in the Oort Comet Cloud (Figure 1.6) that marks the true outer boundary of the solar system, stars in a dense elliptical or irregular cluster within our Milky Way Galaxy (or other galaxies), the spiral-arm structure of the Milky Way Galaxy (or other galaxies), and clusters of galaxies within the universe. The problem is rendered even more difficult when the individual particles are too small to be separately resolved or when the number of particles is so large that a rigorous particle-by-particle formulation is impractical. The four planetary ring systems of our solar system all fall in the more difficult category; they are also the only such arrays we can observe at relatively close range. If our observations of these ring systems can help us to understand their detailed behavior, it is possible that our conclusions about ring systems can be used

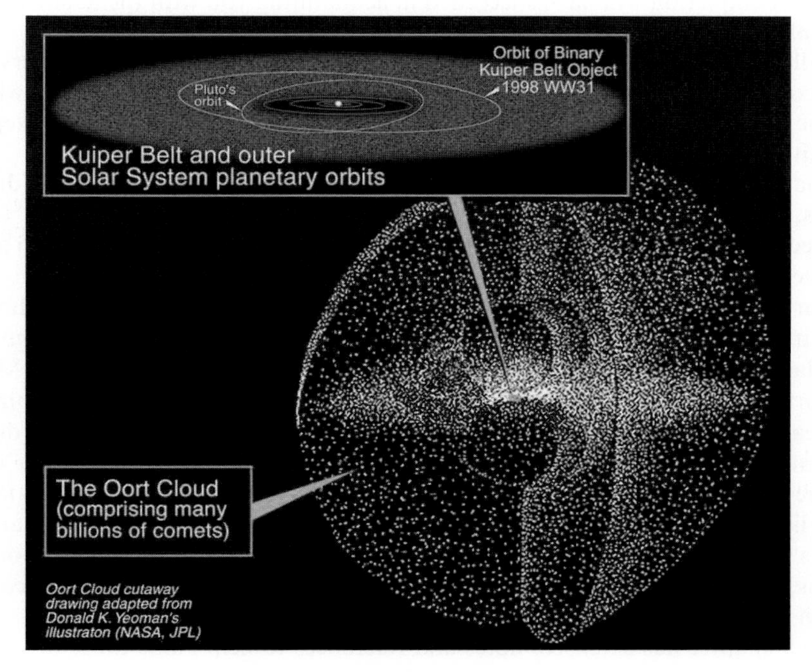


**Figure 1.6.** This cutaway sketch of the spherical Oort Comet Cloud and the much smaller Kuiper Belt disk were rendered by a NASA Jet Propulsion Laboratory's comet research scientist, Donald K. Yeomans. The Sun and all the planets of the solar system are located at the very center of this diagram.

to help us understand their larger counterparts mentioned above. It is in that sense that planetary ring systems serve as a natural laboratory for the studies of many-body interactions elsewhere in the universe.

Some relationships of this nature have already been established. Some gaps in Saturn's rings are caused by the gravitational influence of the satellite Mimas; gaps in the asteroid belt are due to the gravitational influence of the planet Jupiter. While no such relationship has been established between the formation of spiral structure in galaxies and the spiral density waves and bending waves in planetary rings, there is reason to believe that there are similarities in the formative processes. The formation of a relatively flat Kuiper Belt beyond the orbit of Neptune and the flatness of the arms of a spiral galaxy are reminiscent of the flatness of planetary rings as well. Continued study of these and other phenomena may reveal a large variety of characteristics shared by planetary ring systems and their larger cousins.

## 1.9 OUTLINE OF THE REMAINDER OF THIS BOOK

So join us in this journey of discovery and examination of planetary rings. We hope you will find it as fascinating to contemplate as we have, and we trust that the journey's completion will have succeeded in acquainting you with the detailed nature of the planetary ring systems within our solar system.

Following this introductory chapter, Chapter 2 will provide a detailed overview of the discovery and early findings relative to Saturn's rings. Chapters 3, 4, and 5 will deal with the discoveries and early findings on the rings of Uranus, Jupiter, and Neptune, respectively.

Chapters 6, 7, and 8 will summarize all we presently know about the Jupiter, Uranus, and Neptune ring systems, with possible alternative explanations for the observed features in those cases where there is not yet a unanimity of opinion among ring scientists.

Our coverage of the Saturn ring system is extensive enough that we have chosen to split it into two chapters. Chapter 9 will summarize the state of our knowledge of the granddaddy of all ring systems prior to the arrival of the international Cassini Mission at Saturn in July 2004. Chapter 10 will be a progress report on what new things we have learned about the Saturn system from approximately 18 months of detailed observations and analysis of Cassini Orbiter data.

The final two chapters are attempts to provide a reader's digest of the data from the earlier chapters, doing a detailed comparison of the four known planetary ring systems (Chapter 11) and providing a possible roadmap to future observations and analysis, including extension of planetary ring system findings to their larger counterparts in the solar system and beyond (Chapter 12).

Each chapter has extensive notes and references to help guide those of you who would like to refer to original research or have more detailed explanations of certain concepts than can easily be provided in the chapter texts themselves. We have also included a bibliography at the end of each chapter to books or summary articles that have been consulted in the preparation of the chapters or that provide other authors' views on ring systems. Some of you may want access to electronic versions of the pictures or diagrams used in this book. Most are from NASA, and a section at the end of each chapter will provide sources where these are available in relatively high resolution, sometimes in color, as contrasted with the mainly black and white presentations in this book. NASA images are also identified by image number within their respective figure captions.

At the end of the book, we include a detailed index to the contents of the book. The index has been designed to provide a method of finding content material for those who have already read this book but want to refer back to some specific information, or for those who choose not to read this book from beginning to end but are interested in specific portions of the content.

With that semi-brief introduction to the world of planetary ring systems, we invite you to fasten your seatbelts and join us for what promises to be (and indeed already has become) a fascinating journey.

## 1.10 NOTES AND REFERENCES


[1] Maxwell, J. C., 1859, *On the Stability of the Motions of Saturn's Rings*, Cambridge and London: MacMillan and Company. Reprinted in *Scientific Papers of James Clerk Maxwell*, Vol. 1, Cambridge University Press, 1890.

[2] Keeler, J. E., 1895, "Spectroscopic proof of the meteroitic constitution of Saturn's rings", *Astrophysical Journal* **1**, 416–427.

[3] Anderson, John D., 1997, "Gravitation", in *Encyclopedia of Planetary Sciences*, edited by Shirley and Fairbridge, pp. 283–287.

[4] De Pater, Imke, Showalter, Mark R., Burns, Joseph A., Nicholson, Philip D., Liu, Michael C., Hamilton, Douglas P., Graham, James R., 1999, "Keck infrared observations of Jupiter's ring system near Earth's 1997 ring plane crossing", *Icarus* **138**, 214–223.

[5] Showalter, Mark R., 1991, "Visual detection of 1981S13, Saturn's eighteenth satellite, and its role in the Encke gap", *Nature* **351**, 709–713.


## 1.11 BIBLIOGRAPHY


Burns, Joseph A., 1999, "Planetary rings", in *The New Solar System* (Fourth Edition), edited by Beatty, Petersen, and Chaikin, pp. 221–240.

Burns, Joseph A., Hamilton, Douglas P., Showalter, Mark R., 2003, "Bejeweled worlds", *Scientific American* **13**, 74–83 (special edition of *Scientific American*).

Horn, Linda J., 1997, "Planetary ring", in *Encyclopedia of Planetary Sciences*, edited by Shirley and Fairbridge, pp. 602–608.

Spilker, Linda J., editor, 1997, "Those magnificent rings", in *Passage to a Ringed World, The Cassini-Huygens Mission to Saturn and Titan*, NASA SP-533, pp. 41–52.

Time-Life Books, 1991, "Mysterious ring worlds', in *Moons and Rings* (part of a series entitled "Voyage Through the Universe"), pp. 92–133.

## 1.12 PICTURES AND DIAGRAMS

Figure 1.1 *http://photojournal.jpl.nasa.gov/jpeg/PIA03156.jpg*
Figure 1.2 *http://www.physics.byu.edu/Graduate/photo29.jpg*
Figure 1.3 *http://photojournal.jpl.nasa.gov/jpeg/PIA01622.jpg*
Figure 1.4 *http://photojournal.jpl.nasa.gov/jpeg/PIA02207.jpg*
Figure 1.5 *http://pds-rings.seti.org/saturn/vgr2_iss/c4399350.gif*
Figure 1.6 *http://upload.wikimedia.org/wikipedia/en/0/03/Kuiper_oort.jpg*

# 2

# The discovery of the Saturn ring system

## 2.1 FIRST OBSERVATIONS OF SATURN'S COMPANIONS

The invention of the telescope ended the age of naked-eye astronomy. Astronomers no longer needed to scan the heavens with their eyes alone. The "looking glass" opened the solar system and indeed the universe to the scrutiny of humanity. The first recorded use of the telescope for astronomical study was performed in 1610 by Galileo Galilei. Using just a 20-powered spyglass, Galileo discovered spots on the Sun, four satellites about Jupiter, and surface features on the Moon (see Figure 2.1).

His curiosity also led him to turn his attention and his looking glass towards Saturn. In that same year, in the middle of July, Galileo was surprised to see that Saturn was not a single sphere. It appeared to be composed of three individual spheres [1] (see Figure 2.2).

With this extraordinary discovery came an extraordinary problem. Galileo realized that some unscrupulous astronomers might try to take credit for his discovery. Some approach had to be taken to preserve his claim. To safeguard his discovery, he encrypted his results into an anagram and then asked others to come forward with their own observations and explanation [2, 3]. His anagram was as follows:

*Smaismrmilmepoetaleumibunenugttauiras.*

Johannes Kepler (1571–1630), who had been following Galileo's telescopic discoveries with great interest, believed that Galileo's anagram had to do with companions around Mars [4]. Kepler believed that the solar system followed specific laws based on simple mathematical relationships. Since Earth had one moon and Jupiter had four, Kepler firmly believed that Mars should have two. Applied further out into the solar system, this simple arithmetic relationship would predict that Saturn should be surrounded by eight moons. Kepler's belief was really just a guess and nothing more. However, he was so convinced of the correctness of his belief that he tried

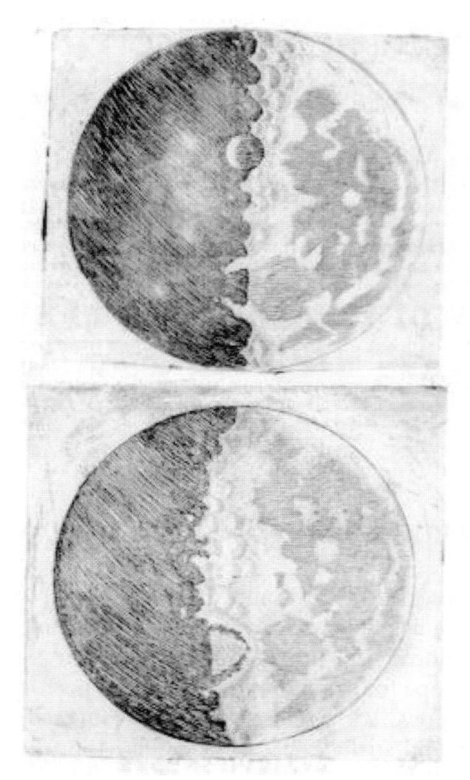

**Figure 2.1.** Detailed drawings of the Moon made by Galileo in December 1610 show lunar surface features.

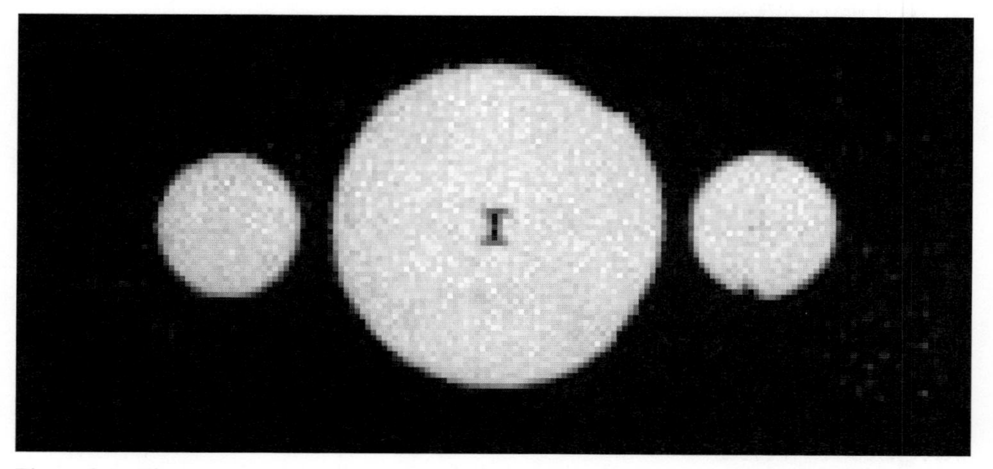

**Figure 2.2.** The poor resolving power of Galileo's telescope prevented him from understanding the true nature of the objects that surround Saturn.

to re-arrange the letters in Galileo's anagram to corroborate his interpretation. From Galileo's clue, Kepler was able to rearrange the letters in the anagram into the following phrase:

*Salue umbistineum geminatum Martia proles.*

The translation of this was "Hail, twin companionship, children of Mars" [5]. However, in November 1610, Galileo published the true solution to his anagram:

*Altissimum planetam tergeminum obseruaui.*

This translation was, "I have observed the highest planet to be triple-bodied" [6]. Galileo believed that the other objects could not be large moons orbiting Saturn. His assertion stemmed from the fact that Jupiter's four moons were solitary points of light and could be observed to orbit about the planet. In contrast, Saturn's companions appeared much larger relative to the planet and appeared motionless.

The static nature of the three bodies made Galileo lose interest in studying the planet. Consequently, his observations of Saturn became less and less frequent. Unbeknown to Galileo, the aspect of the rings was approaching an edge-on orientation. With only an occasional observation of the planet, Galileo was surprised by what he saw in the fall of 1612. To his astonishment, the two companions had vanished. Galileo was mystified and could not explain their disappearance. However, he was confident that they would return. As a matter of fact, he was so certain that he developed a detailed theory about how the bodies had disappeared and how they would return. In the summer of 1613, the companions did reappear, but not in the manner Galileo had predicted.

Over the next three years, few observations of Saturn were made, possibly due to the very slow changes in the system's appearance. Most astronomers lost interest in Saturn and turned their attention to more stimulating astronomical questions. No one recorded the gradual changes of Saturn and its companions between 1613 and 1616. When Galileo again turned his telescope toward Saturn in September 1616, he was again shocked by what he saw. The change in the shape of Saturn's companions prompted Galileo to write:

*"two companions are no longer two small perfectly round globes ... but are at present much larger and no longer round ...* [7]

A drawing made by Galileo in 1616 showed a spherical planet that can be best described by the planet having "cup handles" (see Figure 2.3). To this day, because of this description, the edges of Saturn's rings are often referred to as *ansae*, the Latin word for handles. Unfortunately, the slowly changing shape of Saturn's companions were perceived more as an oddity than as a serious astronomical problem worthy of scientific inquiry. During the next thirty years, Saturn's companions were frequently observed, but very little progress was made toward explaining them. Even the

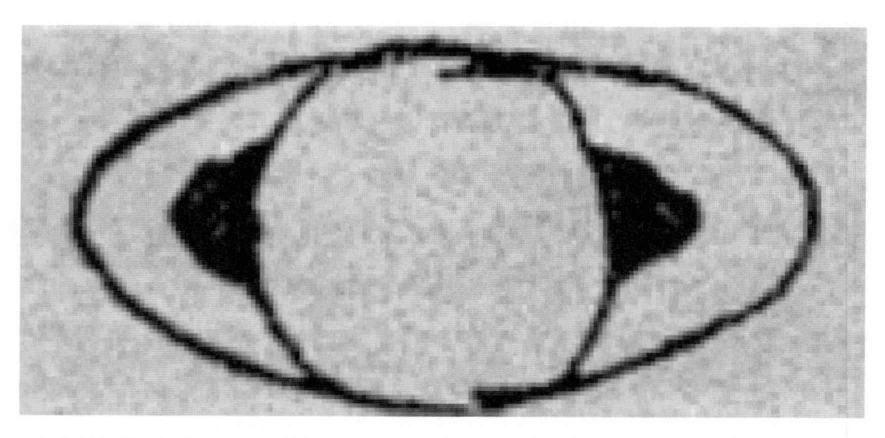

**Figure 2.3.** Galileo's drawing of Saturn as seen in 1616 clearly show objects that appear to look like cup handles. The objects were no longer spheres as first reported in 1610.

disappearance of Saturn's companions in 1626 and their reappearance the following year did little to excite the scientific community of the day.

## 2.2 A MODEL FOR THE RING EMERGES

In 1642 astronomers began once again to take interest in the problem of Saturn's companions and their slowly changing shape. Notable astronomers like Pierre Gassendi (1592–1655), Francesco Fontana (1585–1656), Antonius Maria Schyrlaeus de Rheita (1597–1660) and Johannes Hevelius (1611–1687) all began to make systematic observations of the planet [8]. Although the accumulation of a greater number of observations was necessary for determining the true nature of Saturn and its companions, many of the observations seemed to contradict each other. Scientific progress using apparently conflicting data was difficult at best.

Early in 1655 Christiaan Huygens (1629–1695) began his astronomical observations of Saturn. During that winter he applied the logic established by Rene Descartes (1596–1650) to his planetary observations. This logic allowed Huygens to make the following statements: Saturn was its own system, Saturn should rotate about its own axis, and material located between Saturn and its newly discovered moon Titan should rotate with a period less than that of the moon (estimated to be about 16 days). Because there were no visible changes in the shape of Saturn's companions in 16 days, the companions had to have a shape that was symmetric about the planet's spin axis [10]. In other words, Saturn must have a ring!

Like his fellow astronomers, Huygens wrote his theory on the nature of Saturn's companions as an anagram in *De Saturni Luna Observatio Nova* [11] in March 1656. Though this was written in 1656, most of the astronomical community did not become aware of his theory for another three years when it was published in *Systema Saturnium* (see Figure 2.4). Some members of the astronomical community adopted

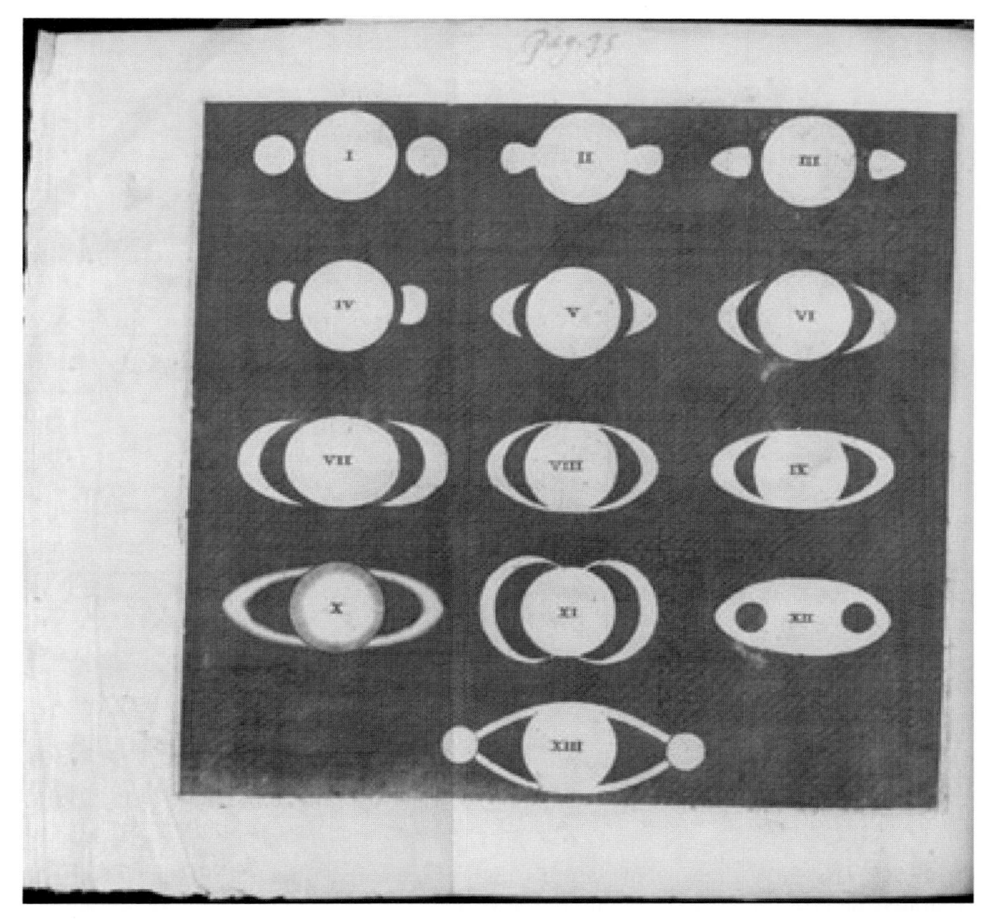


**Figure 2.4.** Drawing from Christiaan Huygens' *Systema Saturnium*, drawn by various Saturn observers between 1610 and 1650.

Huygens' theory immediately, while others did not. Those who were skeptical had problems with Huygens' characterization of the ring. It was difficult for some to believe that the ring was solid, had an observable thickness (approximately 4,000 km), and was inclined to the planet's orbital plane. Even those who were proponents of Huygens' ring model thought the ring was not thick but very thin. Many of these astronomers believed that the ring was so thin that, when viewed edge-on, it would essentially disappear.

However, some astronomers had difficulty believing that Saturn's companions were really the two ansae of one large ring. Johannes Hevelius believed that the two crescents were attached to either side of the planet. Even though he was wrong about the companions, he was able to correctly explain the periodic nature of the planet and its companions. His drawing, made in 1656, contained 24 sketches of crescents about

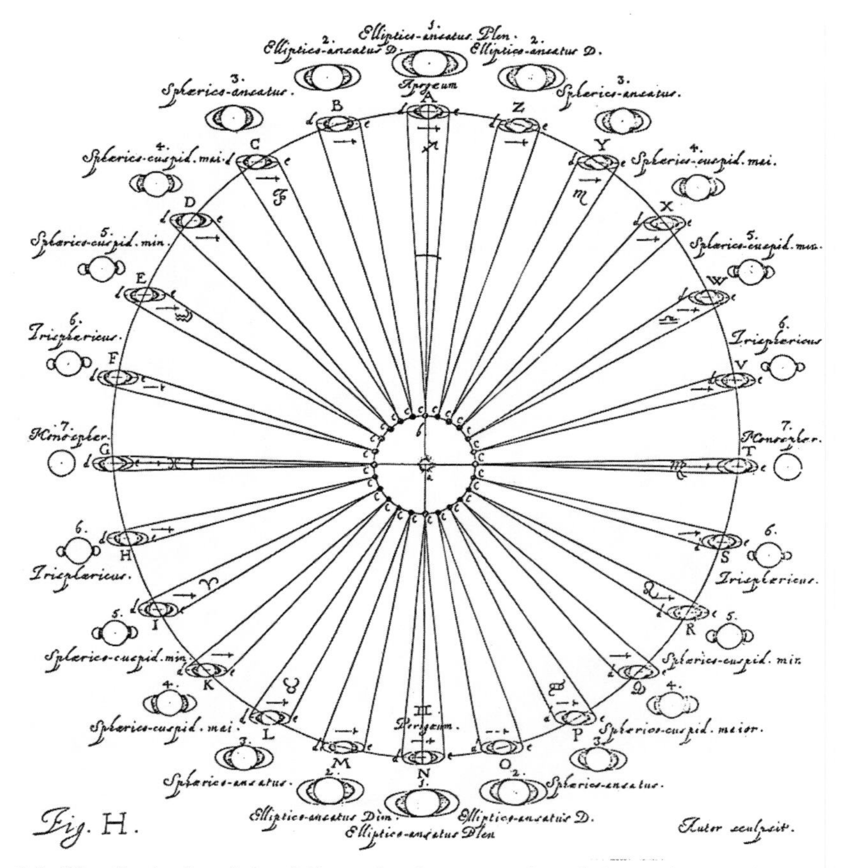


**Figure 2.5.** Hevelius's drawing of Saturn's phases as viewed from the Earth was first to accurately describe the periodic nature of the system.

Saturn and how they would appear from Earth as the planet moved around the Sun (see Figure 2.5).

Two years later, Christopher Wren (1632–1723) developed an alternative to rings and crescents. He claimed that Saturn had an elliptical corona attached to it. This corona was supposedly fluid-like and was emitted from the planet's equator. As the planet and its corona orbited the Sun, the corona would present different aspects as viewed from Earth. Wren believed that the corona would disappear when it was seen edge-on [9]. However, Wren's hypothesis still had the fundamental shortcoming that it could not explain how coronas (or any other shapes) could exist in the heavens.

While aware of these competing hypotheses, Huygens was still convinced that his ring theory was correct, and he continued to refine it. His first challenge was to somehow resolve the issue of the ring's inclination about the planet. From the beginning, Huygens had assumed that Saturn's ring was inclined to its orbital plane by 23.5°, just like Earth's equator was to its orbital plane (see Figure 2.6). The actual

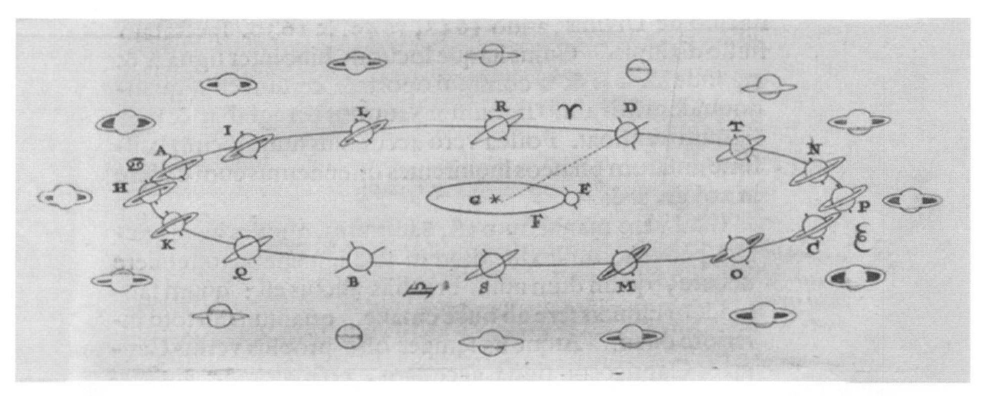


**Figure 2.6.** These drawings were made by Christiaan Huygens and published in 1659 in *Systema Saturnium*. Even though he made his final change to the ring's inclination in 1668, he clearly understood the reasons for the changing appearance of Saturn's rings many years earlier.

inclination of Saturn's ring is about 26.7°, and this error led to problems when Huygens tried to determine the thickness of the ring. It took several years, but in 1668 Huygens made an adjustment to the ring's inclination such that his ring model came reasonably close to predicting the edge-on disappearance of the ring. When the ring disappeared in 1671 as his revised theory predicted, opposition to his theory also disappeared and the reality of Saturn's ring became an accepted scientific fact.

## 2.3 THE NATURE OF THE RINGS

The ring's thickness was not the only erroneous conclusion Huygens had made about the ring: Huygens had also proposed that the ring was a solid body [12]. First, astronomers had difficulties explaining the stability and periodic motions of a solid spinning ring. Remember that astronomers of the day did not yet understand the law of gravitation or the laws of planetary motions. Newton's *Principia Mathematica*, which contained the first formulation of the law of gravitation, would not be published until 1687.

Astronomers came up with two alternate explanations for the nature of the rings, either of which might help to resolve their difficulties with the solid-ring theory. The ring could be composed of a cloud of vapor, or it might be composed of a very large number of small satellites [13]. Scientific literature indicates that the "large number of small satellites" hypothesis was first proposed by the French poet Jean Chapelain (1595–1674) in 1660 [14]. Given a choice between the vapor-cloud and the small-satellite models, most astronomers chose the small satellite model as their favorite.

In 1664 Giuseppe Campani (1635–1715) made an etching of Saturn with an interesting difference from those made by his predecessors and contemporaries. Campani's etching had a bright inner ring with a dimmer outer ring [15]. Twelve years later

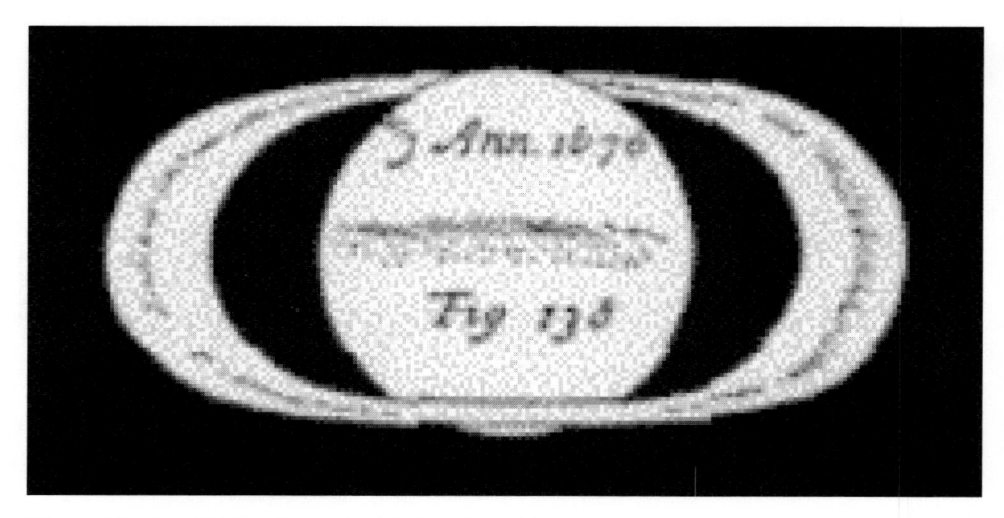

**Figure 2.7.** In 1676 Cassini made this drawing of Saturn with a ring that appears to be divided into an inner and an outer ring.

Giovanni Domenico Cassini (1625–1712), observing Saturn from the Paris Observatory, noticed that the outer dimmer ring and the brighter inner ring appeared to be separated by a gap or division (see Figure 2.7). The idea left astronomers in the awkward position of trying to understand how a planet could have two solid spinning rings. This difficulty added to the popularity of the "large number of very small satellites" model.

Observations continued for the remainder of the 17th Century and into the 18th Century. With each passing decade came better and better data from which to resolve the true nature of Saturn's ring. Many prominent astronomers, including William Herschel (1738–1822), continued for a time to cling to the concept of a solid ring encircling Saturn [16]. However, astronomers kept observing features that made it increasingly difficult to defend a solid ring model. For example, James Short (1710–1768) believed that he had observed multiple gaps in Saturn's ring. It was not until Pierre Simon de Laplace (1749–1827) and his analysis of the ring that astronomers conceded that Saturn's ring could be composed of multiple divisions. Their ring model had many individual, solid, thin rings with spaces between each one [17].

The number of reported gaps and divisions continued to grow. The issue of assigning credit to the gaps in the rings was and still is quite convoluted. The number of gap observations, coupled with lack of precision for describing their position in the rings, has led to many differences of opinion over who was the first to observe a particular gap. Along with gaps, some light and dark features were also reported. In 1837 Johann Franz Encke (1791–1865) recorded a low contrast feature in the middle of the outer ring (see Figure 2.8). In that very same year Encke's assistant Johann Galle (1812–1910) reported that there appeared to be a ring interior to the already known bright inner ring. This proposed inner-most ring was detected by the shadow

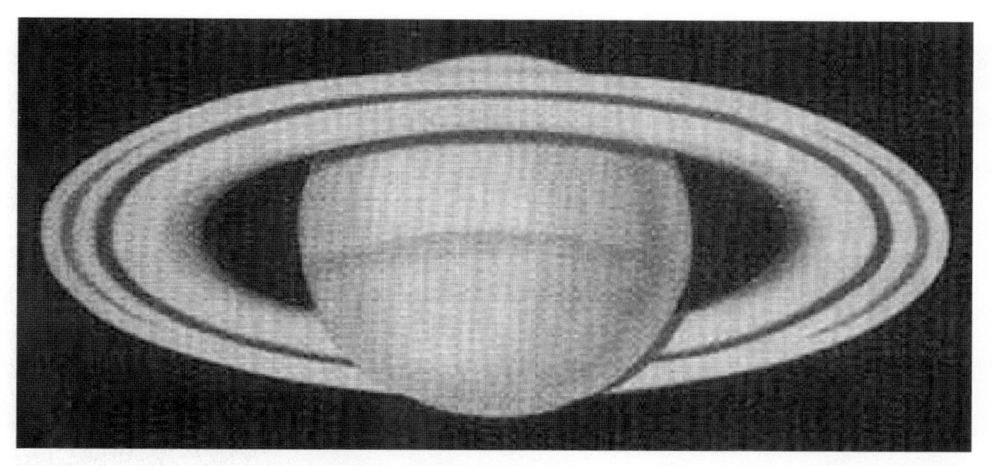

**Figure 2.8.** Drawing by Johann Franz Encke in May of 1837, showing the broad, low-contrast feature in the middle of the A ring that is now called the Encke gap.

that it cast on the atmosphere of Saturn. To standardize a naming scheme and to clarify which portion of the ring astronomers were referring, the outer ring became known as the A ring, the brighter middle ring was called the B ring, while the ring closest to the planet was named the C or Crepe ring.

As telescopes improved, better and better measurements of the radial width of the rings were made. Data from different telescopes made at different times showed that the width of the rings had varied over time. This made it increasingly difficult to hold on to the belief that the rings were solid. By the middle of the nineteenth century Laplace's model for multiple solid rings was appearing less likely.

In 1850, George P. Bond (1825–1865) evaluated Laplace's work and realized that thin solid rings would be unstable [19]. In addition, observations of the translucent C ring lent support to a non-solid ring model. All of these disjointed facts made understanding the true nature of Saturn's ring a major astronomical problem. In hopes of obtaining an answer to this dilemma, the University of Cambridge in 1855 made this problem the subject of the Adams Prize Essay [20]. In 1857 the prize was won by James Clerk Maxwell (1831–1879), who successfully explained all the observations by assuming the ring was composed of an infinite number of small particles.

The last remaining unexplained observations to be incorporated into a Saturn ring model were the many gaps found in the rings. To be successful, a modification to the ring model had to describe the mechanism by which the Cassini Division, the Encke gap, and the numerous other dark ring features were produced (see Figure 2.9).

An explanation for the existence of ring gaps was finally formulated in 1866 by Daniel Kirkwood (1814–1895) with his proposal of ring resonances. Ring resonances occur when the orbital period of one of Saturn's moons is in some integer multiple of the orbital period for a particular region of the ring. In this case, Kirkwood noticed that a particle in the Cassini Division would orbit exactly three times for every orbit of

**Figure 2.9.** Drawing by James Keeler made in January 1888 showing what is now called the Encke gap near the outer edge of the A ring.

Enceladus. At the time, the idea of resonances was still in doubt until Kirkwood was able to use it to successfully explain the structure of Saturn's rings and the gaps in the asteroid belt. Today the gaps in the asteroid belt are known as Kirkwood gaps in his honor.

From the late 1800s to the middle of the 1900s astronomers continued to refine the accepted model of Saturn's ring. The planet had three rings (A, B, and C), each composed of ring particles, and the spatial density (i.e., the number of particles in a volume of one cubic meter) of ring particles in each was responsible for its observed brightness (i.e., the amount of light reflected back towards the Earth). With the advent of photographic plates and larger telescopes, details in Saturn's rings became more and more apparent.

These new details led some astronomers to believe that there might be a ring even fainter than the C ring between the inner edge of the C ring and the top of Saturn's atmosphere. Although the existence of such a ring was never verified from ground-based observations, the letter was "reserved" for such a ring. The D ring since observed by NASA spacecraft could not have been detected from ground-based telescopes.

The last major ground-based ring discovery was made in 1967 by Walter A. Feibelman (1925–2004). Feibelman was working as an assistant research professor of physics and astronomy at the University of Pittsburgh. While studying Saturn images taken by the university's Allegheny Observatory, Feibelman thought he saw material in orbit outside of the planet's main ring. Combining his observations with mathematical calculations, Feibelman believed he had enough data to support the discovery of a new ring [22]. If true, this would be Saturn's E ring. Unfortunately, other ground-based data would not be able to support his claim. Confirmation would require a close-up visit by a robotic ambassador from Earth.

## 2.4 ENCOUNTERS WITH THE RINGS

The birth of the space age in 1957 had a profound impact on the study of planetary rings. For the first time, rings could be observed from the vicinity of the ring system itself. This includes views from behind the planet as seen from the Sun. The advantage of this is that ring particle size can be determined by how light from the Sun is scattered by them. Large particles reflect light back towards the Sun (i.e., the light is back-scattered), while small particles are more visible when viewed from the direction opposite that of the Sun (i.e., the light is forward-scattered). Forward-scattering is particularly noticeable when driving a car with a dusty windshield toward a rising or setting sun. Thus rings composed of small ring particles, as seen from behind, appear bright (see Figure 2.10).

The first spacecraft to visit the Saturnian system was Pioneer 11. Pioneer 11, which was renamed Pioneer Saturn in honor of its historic encounter with the planet, reached its closest approach to the planet on September 1, 1979. Unfortunately for the remote sensing community, Pioneer Saturn was designed primarily as a fields-and-particles spacecraft. The consequence of this was that the spacecraft was spin-stabilized rather than fixed in its orientation in space. Consequently, a framing camera was not included as one of the instruments in Pioneer Saturn's science payload. The visible-light-collecting device that was included was the imaging photopolarimeter (IPP) (i.e., a scanning radiometer) and was not suited for high-resolution imaging.

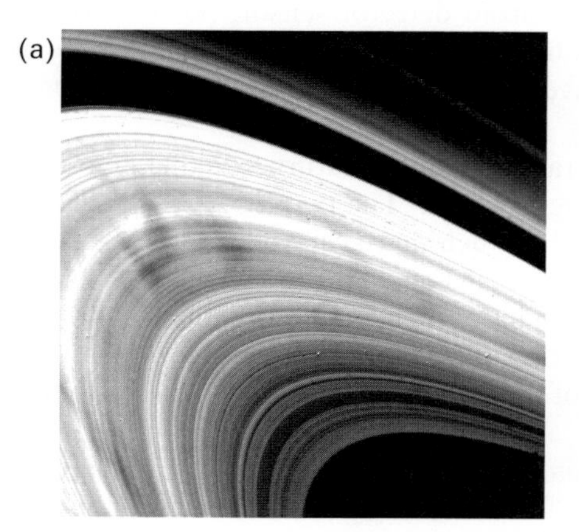


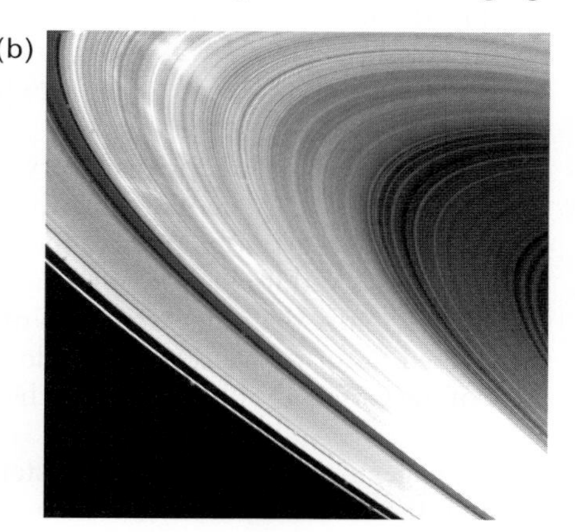


**Figure 2.10.** Radial spokes in Saturn's B ring are composed of tiny micrometer-sized particles. This is evident from this pair of Voyager images, where the spokes appear dark against the B ring in (a), taken at low phase angle by Voyager 2, and bright against the B ring in (b), taken at high phase angle by Voyager 1.

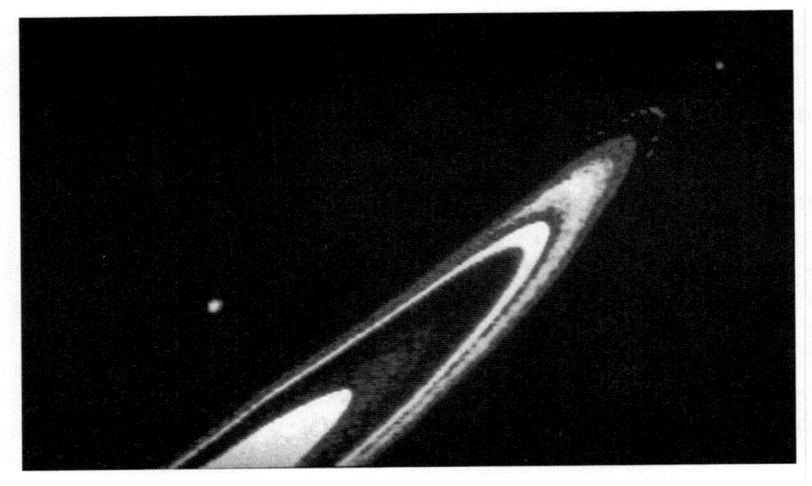

**Figure 2.11.** Pioneer 11 image of Saturn's rings taken on September 1, 1979 from a distance of 941,000 km. The F ring can be seen just outside of the A ring at the upper right in this image. (Pioneer image F-17)

In spite of the instrument's limited capability, Pioneer Saturn discovered Saturn's F and G rings [21]. The F ring was seen as a narrow ring just outside of the A ring (see Figure 2.11). The IPP was only able to detect the densest portion of the F ring, but the data indicated that it had a double-ring structure [23]. Pioneer Saturn also confirmed the existence of Saturn's E ring.

Outside of the main ring system, Pioneer Saturn also observed a very thin ring, now known as the G ring [24]. Alone, the IPP data were not that conclusive, but combined with the spacecraft's asteroid–meteoroid detector sensor, positive identification of the structure was confirmed. The first images of the G ring would have to wait for another spacecraft that would have far superior instruments to the ones onboard Pioneer.

As is the case in robotic planetary exploration, Pioneer Saturn was the prelude to the Voyager 1 and 2 encounters with the ringed planet. The two Voyager spacecraft launched four years after Pioneer Saturn had a much more capable instrument suite and were designed for remote sensing observations. Together the Voyager spacecraft discovered many new characteristics of Saturn's ring system. Among them, Voyager discovered Saturn's *real* D ring inside the C ring. The D ring extends from the inner edge of the C ring about half-way to the top of the planet's atmosphere [25] (see Figure 2.12).

Voyager's sensitive imaging system was also able to discern Saturn's G ring. Though the first indication of this ring came from Pioneer Saturn, the first images of the G ring came from Voyager [26]. The Voyager 1 spacecraft also discovered "braiding" in the F ring (see Figure 2.13). This unique tangled structure was thought to be produced from gravitational interactions between the F ring and two "shepherding" satellites, Prometheus and Pandora [27].

**Figure 2.12.** Voyager 2 took this wide-angle image of Saturn's D ring on August 25, 1981, from a range of 195,400 km. The planet's shadow and fine ring structure can be seen. (P-23967)

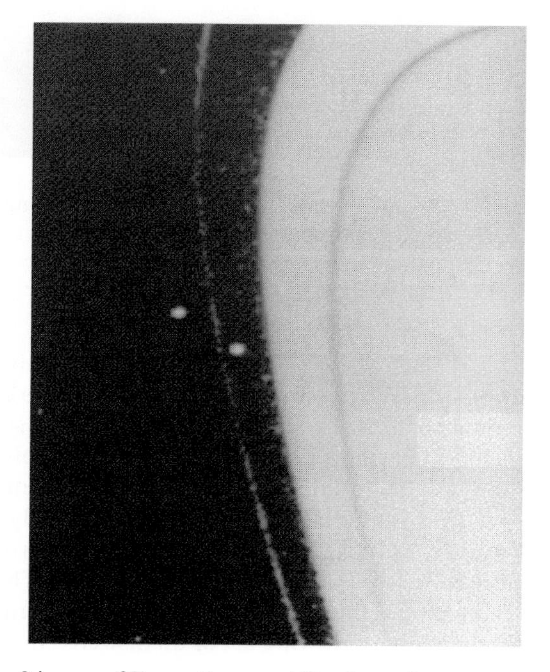

**Figure 2.13.** Voyager 2 image of Prometheus and Pandora, the two moons which orbit Saturn on either side of the F ring and were once thought to confine the F-ring particles. (P-23911)

**Figure 2.14.** Voyager 2 took this high-resolution image of Saturn's rings from a range of 4 million km. Numerous dark "spoke" features in the B ring are seen. Micrometeoroid collisions with B-ring particles may create these spokes, and electromagnetic forces are likely responsible for their maintenance, but no detailed theory has been put forth that adequately explains them. (P-23925)

One of the more curious discoveries by Voyager was the dark radial spoke-like features found in the outer half of the B ring [28] (see Figure 2.14). These features, composed of micrometer-sized particles, revolved about the planet with approximately the same period as the rotation rate of the planet's magnetic field. This may imply that tiny particles in the radial spokes are electrically charged.

By the end of Voyager 2's Post-Encounter phase at Saturn, the planet's ring system was more or less completely defined (see Figure 2.15). The number of rings, their relative optical and physical thicknesses, composition, and radial structure were now known. However, there was a growing list of unanswered questions arising from Voyager data that needed to be answered. These answers would require more time, better ring models and the visit to Saturn by a long-duration orbiter.

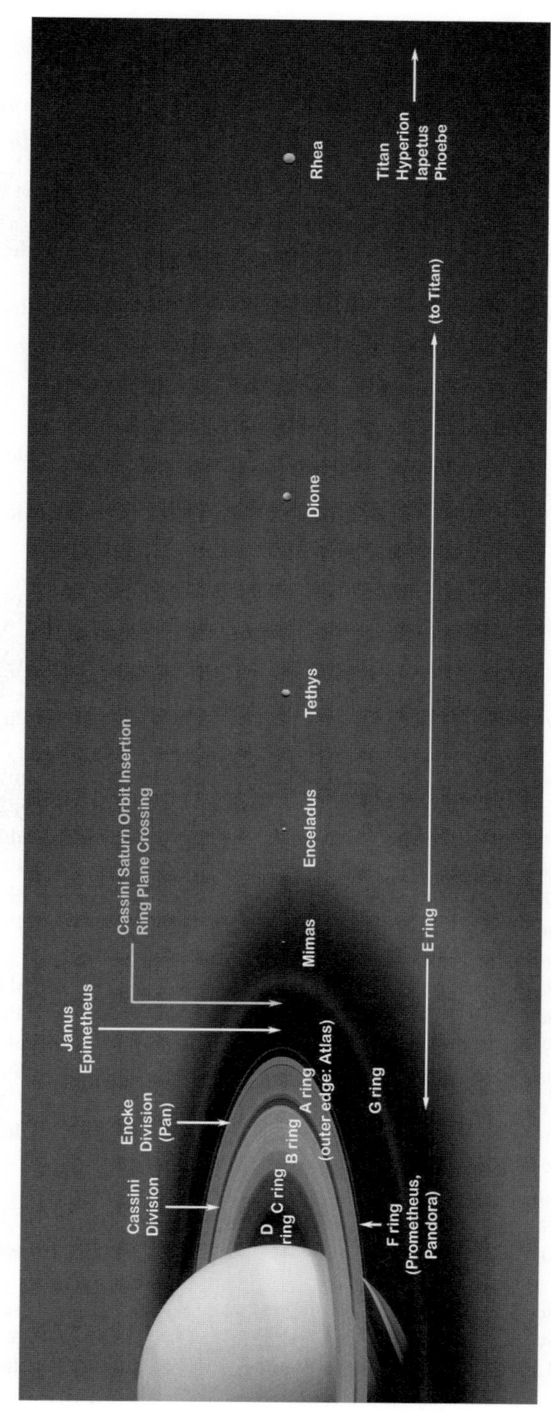


**Figure 2.15.** This artist's concept shows Saturn's seven major rings and some of its moons. Saturn's E ring is the largest planetary ring in our solar system and has a radial extent of approximately 400,000 km. (PIA03550)

## 2.5 NOTES AND REFERENCES


[1] Van Helden, A. (1984) Saturn through the Telescope: A Brief Historical Survey. In: Gehrels, T., Matthews, M. S, editors, *Saturn*, University of Arizona Press, Tucson, pp. 23–43.

[2] Carlos, E. S. (1880) The Sidereal Messenger of Galileo Galilei and a Part of the Preface of Kepler's Dioptrics, London, p. 88.

[3] Galilei, G. (1890–1909) *Le Opere di Galileo Galilei*, Edizione Nazionale, ed. A. Favaro, vol. X, Florence, pp. 409–410.

[4] Kepler, J., *Conversation with Galileo's Sidereal Messenger*, translated by Edward Rosen (New York: Johnson Reprint Co., 1965), pp. 14, 77.

[5] Kepler, *Dioptrice* (Augsburg, 1611); translated in E. S. Carlos, A part of the preface of Kepler's *Dioptrice*, forming a continuation of Galileo's *Sidereal Messenger* (London, 1880; reprint, Dawsons of Pall Mall, 1960), p. 8.

[6] Galilei, G. (1890–1909) *Le Opere di Galileo Galilei*, Edizione Nazionale, ed. A. Favaro, vol. X, Florence, p. 474.

[7] Galilei, G. (1890–1909) *Le Opere di Galileo Galilei*, Edizione Nazionale, ed. A. Favaro, vol. XII, Florence, p. 276.

[8] Hevelius, J. (1647) *Selenographia*, Gdansk, pp. 41–44.

[9] Huygens, C. (1888–1950) *Oeuvres Complètes de Christiaan Huygens*, vol. III, p. 424.

[10] Huygens, C. (1888–1950) *Oeuvres Complètes de Christiaan Huygens*, vol. XV, pp. 294–297.

[11] Huygens, C. (1888–1950) *Oeuvres Complètes de Christiaan Huygens*, vol. XV, pp. 172–177.

[12] Huygens, C. (1888–1950) *Oeuvres Complètes de Christiaan Huygens*, vol. XV, pp. 298–300.

[13] Van Helden, A. (1973) *The Accademia del Cimento and Saturn's Ring*, Physis 15: pp. 237–259.

[14] Huygens, C. (1888–1950) *Oeuvres Complètes de Christiaan Huygens*, vol. III, p. 35.

[15] Huygens, C. (1888–1950) *Oeuvres Complètes de Christiaan Huygens*, vol. V, p. 118.

[16] Van Helden, A. (1984) Saturn through the Telescope: A Brief Historical Survey, p. 32. In Gehrels, T., Matthews, M. S., editors, *Saturn*, University of Arizona Press.

[17] Van Helden, A. (1984) Saturn through the Telescope: A Brief Historical Survey, p. 33. In Gehrels, T., Matthews, M. S., editors, *Saturn*, University of Arizona Press.

[18] *http://members.leapmail.net/~ericj/encke.html*

[19] Van Helden, A. (1984) Saturn through the Telescope: A Brief Historical Survey, p. 35. In Gehrels, T., Matthews, M. S., editors, *Saturn*, University of Arizona Press.

[20] Van Helden, A., (1984) Saturn through the Telescope: A Brief Historical Survey, p. 36. In Gehrels, T., Matthews, M. S., editors, *Saturn*, University of Arizona Press.

[21] Esposito, L. W., Cuzzi, J. N., Holberg, J. B., Marouf, E. A., Tyler, G. L., Porco, C. C., (1984) Saturn's Rings: Structure, Dynamics, and Particle Properties, p. 463. In Gehrels, T., Matthews, M. S., editors, *Saturn*, University of Arizona Press.

[22] *http://lasp.gsfc.nasa.gov/archive/feibelman_obit.html*

[23] Gehrels, T., Bimodality and the Formation of Saturn's Ring Particles, p. 5917. In reprint of the Journal of Geophysical Research, Vol. 85, Number A11, November 1, 1980.

[24] *Ibid.*

[25] Flight Science Office Science and Mission Systems Handbook—Voyager Uranus/Interstellar Mission, Voyager Document PD 618-128 (JPL D-498), Revision C, dated 1985 July 1, Jet Propulsion Laboratory, California Institute of Technology, Pasadena, p. 2.6.

[26] *Ibid.*

[27] *Ibid.*
[28] *Ibid.*

## 2.6 BIBLIOGRAPHY

Flight Science Office Science and Mission Systems Handbook—Voyager Uranus/Interstellar Mission, Voyager Document PD 618-128 (JPL D-498), Revision C, dated 1985 July 1, Jet Propulsion Laboratory, California Institute of Technology, Pasadena.

Gehrels, T., Matthews, M. S., editors (1984) *Saturn*, University of Arizona Press.

## 2.7 PICTURES AND DIAGRAMS

Figure 2.1 *http://www.loc.gov/exhibits/world/images/s75p2.jpg*
Figure 2.2 *http://www.exo.net/~pauld/Saturn/cassinisaturn.html*
Figure 2.3 *http://huygensgcms.gsfc.nasa.gov/Shistory.htm*
Figure 2.4 *http://www.californiasciencecenter.org/Exhibits/AirAndSpace/MissionToThePlanets/Cassini/CassiniUpdates/images/SIL4_9_04.jpg*
Figure 2.5 *planety.astro.cz/saturn/obr/hevelius_phases.gif*
Figure 2.6 *http://www.sil.si.edu/DigitalCollections/HST/Huygens/huygens-guide.htm* (modified by Robbii Wessen)
Figure 2.7 *http://www.pparc.ac.uk/Ed/ch/Rings/RingsMain.htm*
Figure 2.8 *http://members.leapmail.net/~ericj/encke.html*
Figure 2.9 *http://members.leapmail.net/~ericj/encke.html*
Figure 2.10 *http://photojournal.jpl.nasa.gov/jpeg/PIA02275.jpg and http://photojournal.jpl.nasa.gov/jpeg/PIA02269.jpg*
Figure 2.11 *http://spaceprojects.arc.nasa.gov/Space_Projects/pioneer/PNimgs/f17.gif*
Figure 2.12 *http://photojournal.jpl.nasa.gov*
Figure 2.13 *http://www2.jpl.nasa.gov/saturn/gif/saturn92.gif*
Figure 2.14 *http://photojournal.jpl.nasa.gov/jpeg/PIA02275.jpg*
Figure 2.15 *http://photojournal.jpl.nasa.gov/jpeg/PIA03550.jpg*

# 3

# The discovery of the Uranus ring system

## 3.1 FIRST OBSERVATIONS OF URANUS'S RING

It was going to be a long flight, a flight that would leave Perth, Australia on the evening of March 10, 1977, and then, more than 10 hours later, return where it started. This was no ordinary flight and no ordinary plane. The flight was scheduled to leave Perth International Airport and head southwest over the south Pacific Ocean. The goal was to deliver a 0.9-m infrared telescope to an altitude of 12,500 m. At that height the telescope would be above most of the infrared-blocking atmospheric water vapor. The mission objective was to observe the first every predicted occultation of a star by the planet Uranus [1].

The plane was a C-141 and was NASA's Gerard P. Kuiper Airborne Observatory (KAO) (see Figure 3.1). This plane was specifically designed for airborne astronomical observing runs. However, this observing run was unique. This was the first time people knew in advance that Uranus was going to pass in front of a star (SAO 158687), and astronomers would be able to take advantage of it. This star was also a good target, since it was a 9th-magnitude star slightly redder than our Sun, and at infrared wavelengths it would appear as bright as Uranus [2]. If all went as planned, astronomers would be able to make detailed temperature measurements of the planet's upper atmosphere and a precise measurement of its diameter.

There was just one catch. Even if the instrumentation worked perfectly and the atmosphere cooperated, there was still a 17% chance that during the occultation the KAO would be just north of the occultation's northern edge [3]. This made Jim Elliot, the lead astronomer for this observing run, very nervous. It's a very time-consuming, expensive, and difficult endeavor to plan for and execute an observation campaign using an observatory, especially a small one with wings. A science team had to be assembled with mission directors, telescope operators, individuals to run the recording equipment, meteorologists to monitor the atmospheric conditions, and of course the pilots. It didn't help that before he left for Australia, one of his colleagues said

**Figure 3.1.** The Kuiper Airborne Observatory in flight. Notice the telescope opening in the fuselage just aft of the cockpit. The KAO was decommissioned in 1995.

something that would echo in his ears for the duration of the flight. Joe Veverka, a fellow astronomer at Cornell University said, "that a near miss would place an upper limit on a possible Uranian ring system" [4].

The discovery of a ring system around Uranus would be a triumph for observational astronomy, but every grade school pupil knew that Saturn was the only planet in the solar system with rings. Nonetheless, the fact that they could miss the occultation altogether was a frightening thought. It didn't help that a near miss could be used to limit the extent of a hypothetical Uranian ring. After all, Uranus didn't have rings. It became an inside joke for all those on the flight that they were using this occultation event to probe Uranus for rings.

Regardless of the slight possibility of failure, plans were made to perform a coordinated observation with another team. Jim Elliot joined forces with Robert L. Millis, observing from the Perth Observatory in western Australia, to guard against weather problems and equipment errors. A second independent observation of the same event could also be used to corroborate data obtained during their flight. This approach had worked well before, when these two teams performed joint observations of an occultation by Mars in 1976 [5].

The KAO took off from Perth at 10:37 p.m. About five and a half hours later the science team began setting the equipment up for the event. First, they centered the telescope on the target. About 40 minutes prior to the start of the occultation the recorders were turned on and set for continuous data collection. As the time for the occultation approached, the team watched the photometer's signal levels. Then Ted Dunham, Jim Elliot's graduate student who was operating the data-gathering equipment, noticed a dip in the signal strength. The team's first reaction was that the plane had passed through a faint wisp of a cloud that had absorbed the very photons they were trying to measure. But independent measurements indicated that water vapor levels were very low. Then a second dip occurred. At that point Jim Elliot said, "Well maybe this is a D ring. The D ring of Uranus!" [6]. With that everyone laughed. But clearly the team had seen something, and no one was sure what it was.

Then the occultation began. Though this was the event that they had prepared for, the team couldn't help think about the dips prior to the occultation. Would brief dropouts be seen again when the star reappeared from behind Uranus? If they were, had they observed rings, a string of satellites orbiting the planet, or something else? They just had to wait for the star to emerge from behind the planet. When the star reappeared so did the dips in the star's intensity. Something was clearly blocking the starlight, but whatever it was it sure wasn't rings like those seen on the only other known ringed planet. Saturn's rings were very broad, while these events indicated something very narrow. The team knew they had discovered something; they just had to figure out what it was. At this point Jim Elliot began to compose a note to alert the astronomical community about their discovery (see Figure 3.2). Astronomers around the world had to be notified about these unexpected events. Anyone observing this occultation would otherwise turn off their recorders a few minutes after the star emerged from behind the planet. There would be no reason to keep recording unless there was a need to confirm the data dropouts found by the KAO team.

Meanwhile, back on the ground, Robert Millis's team also saw the brightness variations in their occultation data. And they too wondered if the dropouts were caused by clouds, equipment problems, or rings. Unfortunately, their Perth Observatory was further east than the KAO flight path and was only able to record the occultation ingress. Without seeing similar dropouts after occultation egress, the team's interpretation of the intensity variations would naturally tend toward equipment problems, clouds, or highly unlikely occultations by new satellites circling Uranus.

In total, there were five other observatories recording the event: Tokyo, Japan; Peking, China; Naini Tal and Kavalur, India; and Cape Town, South Africa. However, with the exception of the Cape Town Observatory, their latitudes were much farther north than either the KAO flight path or the Perth Observatory. None of the observatories other than the KAO saw the occultation by the planet. While the Perth and Peking Observatories saw starlight dropouts on both sides, they observed only a few (three to five) of the dropouts before Uranus dipped beneath their western horizons. Cape Town Observatory recorded the last of the outbound dropouts, but the inbound and planetary portions of the occultation occurred before Uranus was high enough in the east. Only the KAO recovered the complete occultation data set.

On returning to their hotel in Perth, the KAO team unrolled the recorded data and examined the data dropouts. Rings were briefly considered, but because these data dropouts were so unlike anything expected from rings as they knew them, that possibility was discarded. Whatever caused the dropouts was simply too thin to be rings. With all the activity the night before and the lack of sleep, the tired team

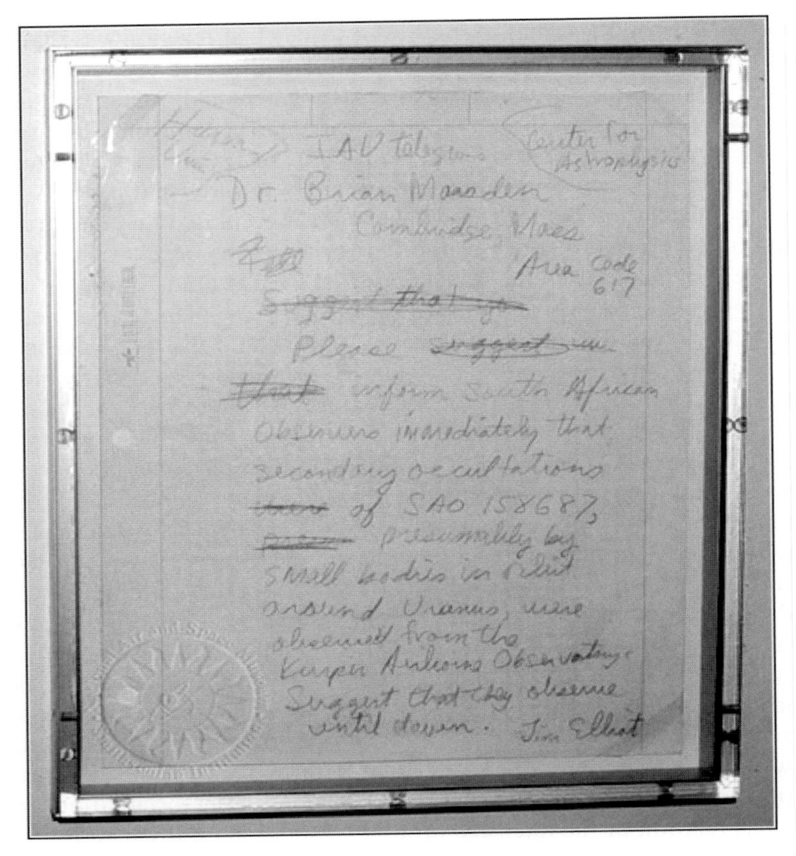
IAU telegram
Center for Astrophysics
Dr. Brian Marsden
Cambridge, Mass
Area code 617
~~Suggest that~~
Please ~~suggest~~
~~that~~ inform South African
observers immediately that
secondary occultations
of SAO 158687,
presumably by
small bodies in orbit
around Uranus, were
observed from the
Kuiper Airborne Observatory.
Suggest that they observe
until dawn. Jim Elliot

**Figure 3.2.** Discovery telegram written by James L. Elliott notifying the International Astronomical Union (IAU) of the discovery of small objects in orbit about Uranus.

considered other possible scenarios, things like an asteroid belt or a belt of satellites around Uranus. But Jim Elliot wasn't convinced. He and his team returned to the United States on March 13, 1977, still questioning the nature of the data dropouts. He finally convinced himself that the planet did have rings the following night when he again spread out the data on his living room floor. There in plain sight were five secondary occultations on either side of the primary Uranus occultation.

Elliot and his team published their results in the science magazine *Nature* [7]. In it the team named the rings after the Greek alphabet. The innermost ring was named Alpha, followed by Beta, Gamma, Delta and Epsilon rings. The Epsilon ring by far was the widest of the rings, but they all were very narrow. However, data from Millis and his team could not confirm all of the same events. The Millis team saw six statistically significant brightness variations, but only three corresponded to those seen by Elliot's team. When the two data sets were compared, Millis's rings 1, 2, and 3 corresponded with Elliot's Epsilon, Gamma, and Beta rings. However, rings 4, 5, and

6 did not correspond to any of the rings reported by Elliot's team. Upon re-examination of their data, Elliot's team found the additional rings reported by Millis's team, along with another ring between the Beta and Gamma rings. This ring they called the Eta ring. Eventually the two different nomenclature schemes were combined to produce the naming scheme we have today.

## 3.2 OBSERVATIONS PRIOR TO VOYAGER'S URANUS ENCOUNTER

The first ground-based images of Uranus's rings, following their discovery, were obtained by Matthews, Neugebauer, and Nicholson at the Palomar Observatory in May 1978 [8]. These images were not able to actually resolve the rings, but they were able to detect them. The technique was to take two images of the planet at different infrared wavelengths. One was taken at 2.2 micrometers, which made the planet appear slightly darker than the rings, and the other at 1.6 micrometers, which made the planet appear much brighter. By subtracting the 1.6-micrometer image from the 2.2-micrometer image, the image shown in Figure 3.3 was produced with the planet virtually removed [9].

Though this technique was able to detect the rings, it clearly was unsatisfactory for determining any of their physical properties. The darkness of the ring material combined with the very low light levels to make Uranus ring images very difficult to obtain. However, this difficulty did little to stop many observers from trying.

One of the more successful observations was done by Rich Terrile and Brad Smith at the 2.5-m du Pont Telescope at Las Campanas Observatory in Chile. Their approach was to observe the planet in 11 different wavelengths from 0.435 to 1.000 micrometer. The multiple wavelength approach was designed to reveal planetary atmospheric features. However, at 0.89 micrometers atmospheric methane absorbs sunlight making the planet appear very dark. Observations taken at this wavelength can be processed to reveal the planet's rings, as seen in Figure 3.4. As was obvious from the stellar occultation observations, the Uranian ring system was dominated by the much more pronounced Epsilon ring than by any of the others. Computer processing gave the image a 3-D feel that has the Epsilon ring appear as a "step" surrounding the planet. Notice that the five major satellites of Uranus can also be seen in the image.

Direct imaging was not the only technique for understanding the structure of the Uranian rings. Stellar ring occultations, like those that led to their discovery, can also be used. However, this time—rather than discovering rings by accident—one can optimize the observation to product better ring results. And indeed this was the case. More than 15 stellar occultations were performed between the rings' discovery in 1977 and the Voyager 2 encounter in 1986. These observations were able to provide better information on the rings' size, eccentricity, inclination, width, and optical thickness. Results of these occultation observations may be found in many books, journals and papers (Elliot [10]; Elliot and Nicholson [11]; French, Elliot, and Levine [12]; and French *et al.* [13]).

Our understanding of Uranus's nine rings continues to improve. The rings, as stated previously, were very different from those around Saturn. Saturn's rings were

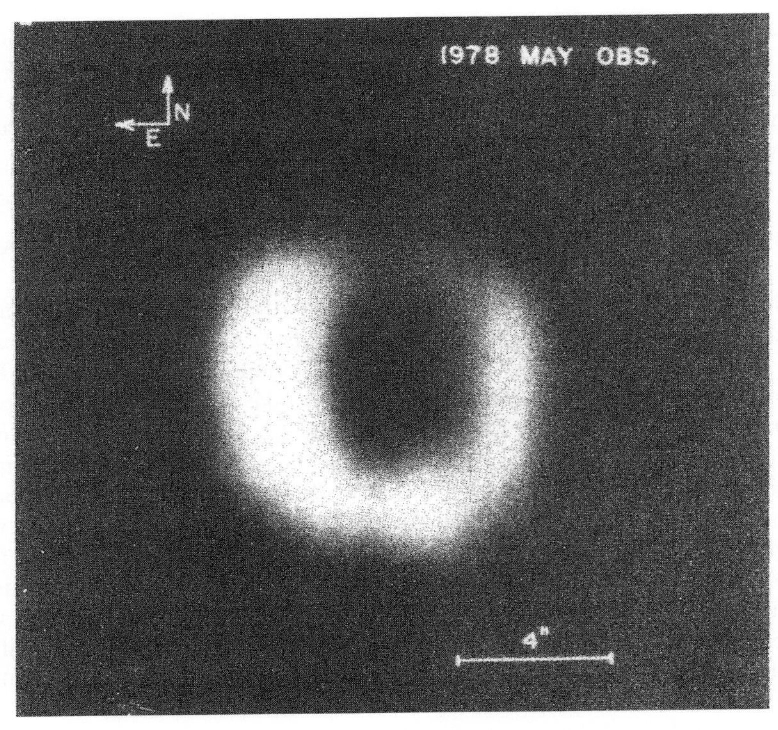


**Figure 3.3.** First ground-based observations of the rings of Uranus. Notice that the planet has all but disappeared from the image while the rings are detected but not resolved. (From Matthews *et al.* [8])

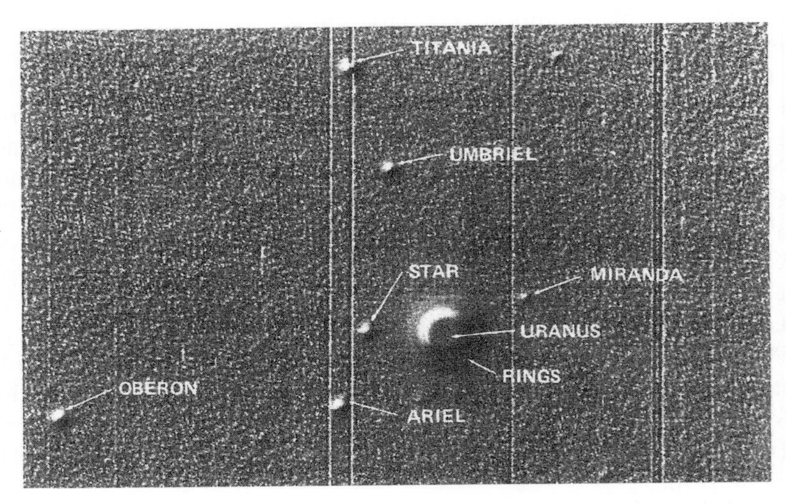


**Figure 3.4.** The textured feel of this image came from the computer-processing of this CCD image. Notice that there still is virtually no observable ring structure. (P-27796)

very wide, whereas Uranus's rings were extremely narrow. In 1979 Goldreich and Tremaine proposed an explanation for the narrowness of the Uranian rings [14]. Their theory suggested that a ring would stay narrow if satellites straddled either side of it. These "shepherding satellites" would force a ring to keep a narrow shape rather than follow their natural tendency to disperse.

A vivid confirmation of their theory seemingly came from analysis of Voyager Saturn ring data. Researchers found two satellites, Prometheus and Pandora, on either side of Saturn's F ring (see Figure 2.13). The F ring's braided behavior seemed to illustrate this tension between the ring's natural tendency to disperse and the satellites' gravitational interactions to keep it narrow.

## 3.3 VOYAGER 2'S ENCOUNTER WITH URANUS'S RINGS

The Voyager 2 Uranus encounter began on November 4, 1985, after a very long journey from Saturn. No one was sure if Voyager would survive this trek, because the craft's design was optimized for encounters with only Jupiter and Saturn. Uranus was well over a billion kilometers and another four and a half years away. In addition, the spacecraft was showing signs of age. The scan platform for pointing the science instruments had a failing azimuth actuator; one of the spacecraft's receivers was dead and the other was "tone deaf"; and the infrared interferometer spectrometer and radiometer (IRIS) and the photopolarimeter (PPS) instruments were showing signs of age, both partially disabled and losing sensitivity. There were also a number of other smaller problems that limited the performance of the spacecraft.

For the Uranus encounter, Voyager science was divided between atmosphere, magnetosphere, ring, and satellite objectives. From ground-based ring observations, scientist knew that it would be particularly difficult to obtain useful ring observations, especially on the inbound leg. These observations were difficult because the rings were narrow and had a very low reflectivity that was only compounded by the extremely low light levels. With all of these challenges and spacecraft problems, JPL engineers had done a phenomenal job nursing the craft to Uranus. This effort paid off with an historic Uranus encounter, and by the end of November 1985 scientists were rewarded with the first detection of the Epsilon ring. This detection occurred with Voyager still 72,300,000 km away from its planetary closest approach [15].

About one day away from its planetary rendezvous, Voyager 2 discovered a new ring between the Delta and the Epsilon rings. This ring was given a provisional name of 1986U1R. This naming convention designated the year of discovery, the planetary body, and that it was the first ring found in that year. However there was more in store for the scientists than just one new ring. About one hour away from closest approach, Voyager 2 performed another set of ring observations. Although badly smeared from the relatively high velocities, one of these frames discovered a second new ring (1986U2R). This ring was different from the other Uranian rings. It was 2,500 km wide and located planetward of the 6 ring. Approximately fifteen minutes later Voyager passed the equatorial plane of the planet. This was followed a few minutes

**Figure 3.5.** Silhouetted image of the nine originally known Uranian rings taken against the planet. This narrow angle image was shuttered 27 minutes before Voyager 2's closest approach. The Epsilon ring is on the right. (PIA01985)

later by an observation that caught the nine original thin rings against the brighter atmosphere of the planet (Figure 3.5).

As previously stated, imaging the rings was very difficult. The rings were just too narrow and dark to obtain much information. A more effective approach was to apply the same technique that was used to discover them nine years earlier. Both the PPS and the ultraviolet spectrometer (UVS) performed stellar ring occultations. These measurements were designed to watch a bright star as the planet's rings passed between it and the spacecraft.

The first stellar ring occultation was of sigma Sagittarii, but the observing geometry was less than ideal. The apparent motion of the star through the ring system just grazed the outer portion of the rings. However, it was able to obtain optical thickness information on the Epsilon, 1986U1R, and the Delta rings. This occultation provided a radial ring resolution of about 10 m [16].

The other stellar ring occultation was accomplished using the star beta Persei. This occultation was a radial cut which crossed all of the known rings, but with a degraded radial resolution of 100 m [17]. Nature appears to have a sense of humor by providing either a good geometry or a good resolution, but not both. A perpendicular pass through the rings results in a relatively shorter time in each ring. The shorter duration produces a lower ring resolution. Fortunately, scientists did not have to make the difficult choice between the two. Mission planners were limited to the viewing geometry provided by the trajectory through the system and by the availability of bright and stable stars. Scientists had no real other choice than to accept the most favorable stars that Nature provided.

Voyager had another approach for obtaining ring information. Rather than using a star as the light source, the spacecraft could transmit its own carrier radio frequency through the rings. Data collected by the Deep Space Network antennas could then be interpreted to provide an understanding of the ring structure and its average particle size. The longer radio wavelengths, when compared with the ultraviolet wavelengths received by the other two instruments, provided another measure of the ring particle size. The radio science subsystem (RSS) ring occultation experiment was able to detect all of the nine previous rings but was unable to see either of the new ones. This placed a limit on the ring particle sizes in these new rings. If the particles were smaller than the RSS wavelengths, then the radio signals would pass through them undiminished in intensity and unaltered in their coherent structure.

There was also the matter of the thin rings and the mechanism for keeping them narrow. As stated previously, the leading theory was that each ring should have small moons on either side of them to maintain their shape. Voyager 2 did discover ten new satellites of Uranus, but only two were close enough to a ring to be considered shepherding satellites. The new shepherding satellites were eventually named Cordelia and Ophelia and were found on either side of the Epsilon ring. Unfortunately, no other satellites were found close enough to the other rings to explain their narrow widths. This could mean that shepherding satellites around the other rings were just too small to be detected or that some other mechanism was keeping them narrow.

On the outbound leg of the Voyager encounter more than 200 high-phase angle ring images were taken to determine the ring's forward-scattering properties. Virtually all of these images were completely black because the rings contained less dust than predicted. However, one imaging frame, the one used to protect the imaging campaign against extreme ring conditions, had both a long enough exposure (96 s) and was taken at a high enough phase angle (172.5°) to provide the most revealing view of the rings (Figure 3.6). This frame, the only one of its kind, shows a dramatic ring structure, and

**Figure 3.6.** The only successful Voyager 2 high-phase angle image of the rings. The long 96-s exposure produced a blurred star field. (PIA00142)

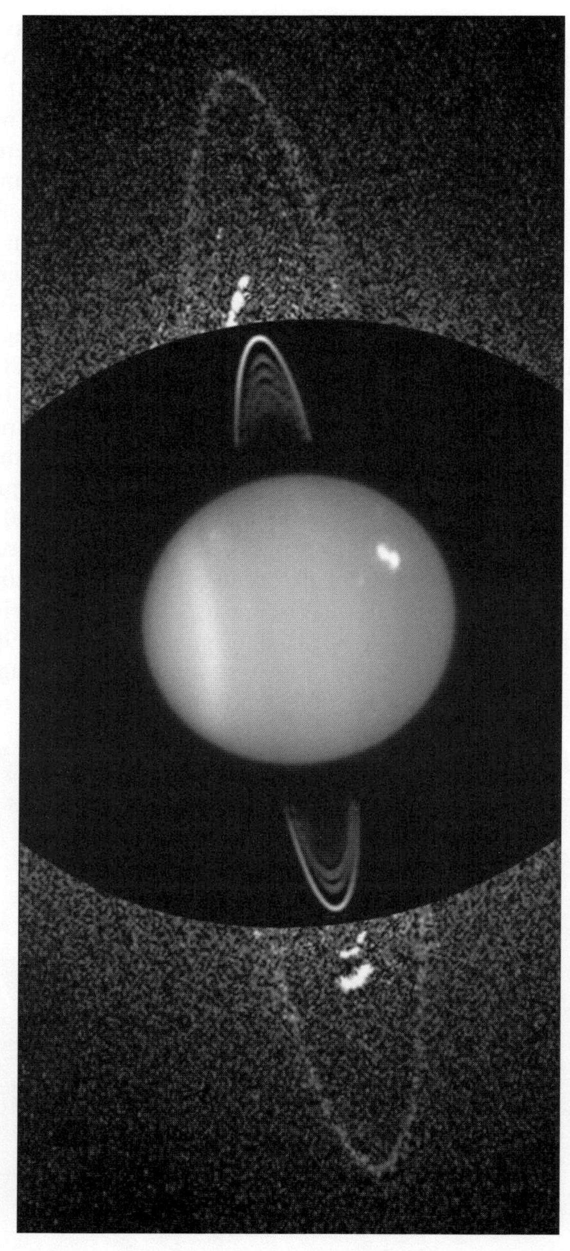

**Figure 3.7.** This composite image of Uranus is made from three different Hubble Space Telescope images. The outer most frame shows the newly discovered pair of outer rings; superimposed on this frame is a long exposure image to reveal the previously known Epsilon ring; and finally an inner frame which shows Uranus with atmospheric features. Courtesy Space Telescope Science Institute (STScI-2005-33)

it will not be duplicated until another spacecraft journeys to Uranus. Unfortunately, since no Uranus mission is currently being planned, this event is at least many decades away.

## 3.4 POST-VOYAGER OBSERVATIONS OF URANUS'S RINGS

Voyager 2's encounter with Uranus was a spectacular success, but it didn't discover all of the planet's secrets. In August 2004 the Hubble Space Telescope, with its Advanced Camera for Surveys instrument, discovered two new moons and two new rings around Uranus (Figure 3.7).

The orbits of the two new rings are extremely far from the planet. The more distant of the two rings (2003U1R) is almost twice the radial distance of the Epsilon ring, the most distant of the previously known rings. The inner of the two new rings (2003U2R) has a radius of 67,300 km and is bounded by the satellite Portia on the inside and Rosalind on the outside [18]. The new outer ring has a radius of 97,700 km and has the newly discovered satellite Mab, orbiting near its center [19] (see Figure 3.8).

On December 22, 2005, Imke De Pater, Heidi B. Hammel, and Seran Gibbard released an image of these two newly discovered rings. The team acquired it on the Near Infrared Camera (NIRC2) mounted on the 10-m Keck II Telescope on Mauna Kea, Hawaii (Figure 3.9). The team had to combine 30 (2-min exposure) images, equivalent to a single one-hour exposure, to make the rings visible [20].

Notice that in this infrared observation the outer of the two new rings could not be seen. Because it was observed at visible wavelengths by the Hubble Space Telescope but not in the infrared, the data suggest that there is a color difference between the two rings. The ring particles of the outer ring, invisible at infrared wavelengths, must be bluer and therefore composed of smaller particles than those of the inner ring. This conclusion suggests different origins for the two rings and could be explained by a hypothesis put forth by M. R. Showalter and J. J. Lissauer [21]. Their hypothesis states that small-to-modest-sized ring particles collide with each other producing smaller particles than the parent particles, but still relatively large. This debris produces a ring that appears slightly red and would have a dusty characteristic. Examples of other dusty rings are Jupiter's rings and Saturn's G ring. The outer ring, which is devoid of these relatively larger particles, is created by a completely different mechanism. In this case, Uranus's new outer ring could be produced by meteoroid impacts with Mab, presumed to be a body largely composed of water ice. The collisions eject icy particles from the satellite's surface to produce a ring that would appear bluer than the relatively larger particle sized inner ring. An example of this type of ring, one that is composed of fine particles, is Saturn's E ring.

If it weren't for the planet's axial tilt, the very faint nature of the rings and the fact that there are no planned future missions to Uranus would mean that very little new information would be gained about these rings in the near future. Fortunately, Uranus is approaching its equinox and its rings will appear edge-on as seen from Earth. This orientation will occur in 2007 and is our best hope for learning more about Uranus's rings.

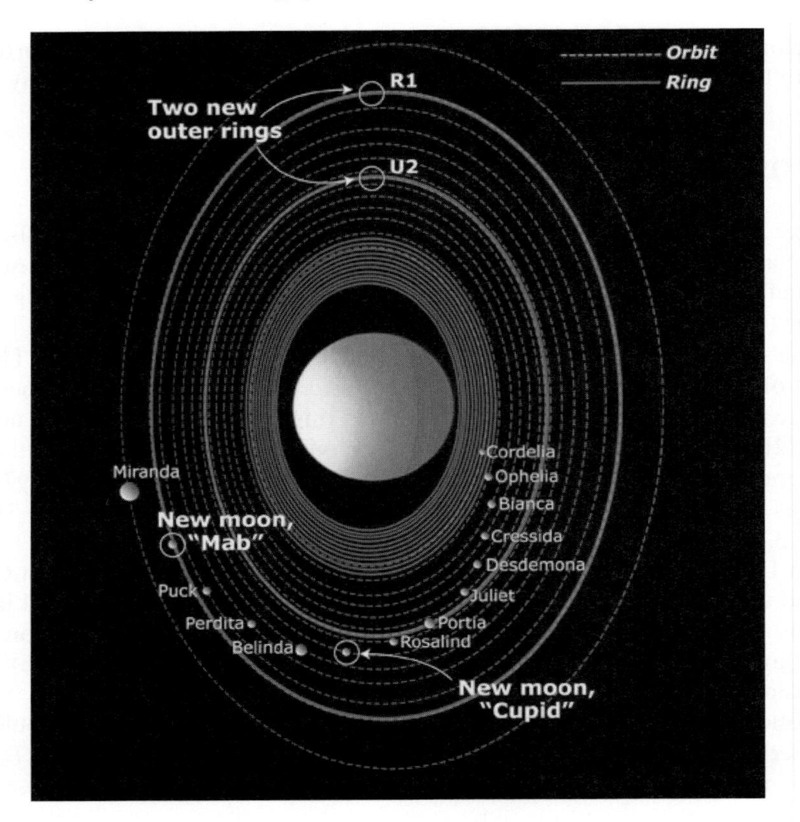


**Figure 3.8.** Artist drawing of the Uranus system with the two new rings and two satellites discovered by the Hubble Space Telescope. The new outer rings are so far from the original rings that they are referred to as Uranus's "second ring system."

## 3.5 NOTES AND REFERENCES

[1] Elliot, J., Kerr, R., 1984, *Rings—Discoveries from Galileo to Voyager*, The Massachusetts Institute of Technology Press, p. 1.

[2] Miner, E. D., 1990, *Uranus: the Planet, Rings and Satellites* (2nd Edition), Wiley-Praxis Series in Astronomy and Astrophysics, p. 54.

[3] Elliot, J., Kerr, R., 1984, *Rings—Discoveries from Galileo to Voyager*, The Massachusetts Institute of Technology Press, p. 5.

[4] *Ibid.*

[5] Miner, E. D., 1990, *Uranus: The Planet, Rings and Satellites* (2nd Edition), Wiley-Praxis Series in Astronomy and Astrophysics, p. 54.

[6] Elliot, J., Kerr, R., 1984, *Rings—Discoveries from Galileo to Voyager*, The Massachusetts Institute of Technology Press, p. 12.

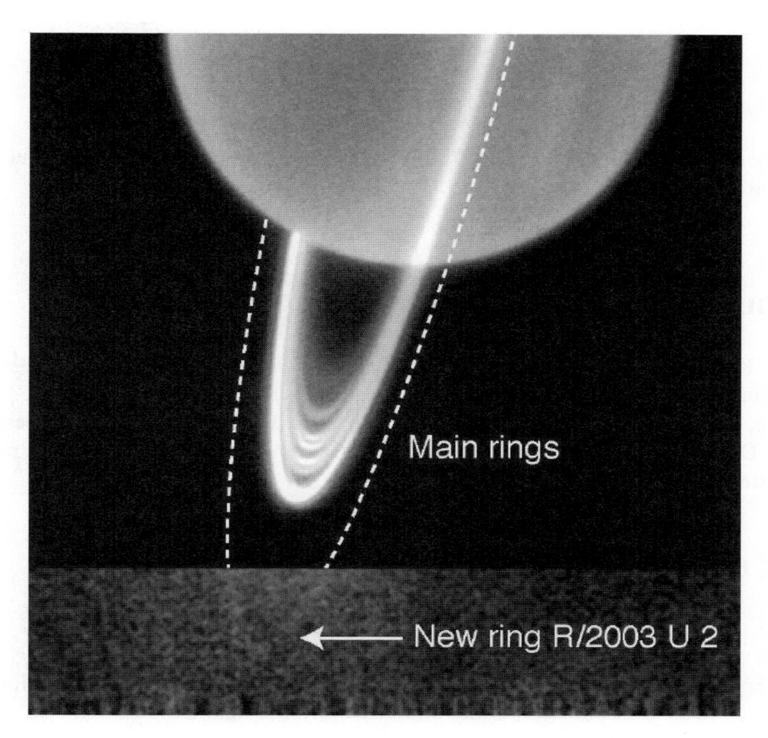


**Figure 3.9.** The dashed line marks the position of the innermost of the two new Uranian rings as compared with its main ring system. Notice that the smaller enhanced inset image does not show the outer ring. (Courtesy of UC Berkeley/SETI/Keck)

[7] Elliot, J. L., Dunham, E. W., Millis, R. L., 1977, "Discovering the rings of Uranus", *Nature* **267**, 328–330.
[8] Matthews, K., Neugebauer, G., Nicholson, P. D., 1982, "Maps of the rings of Uranus at a wavelength of 2.2 microns", *Icarus* **52**, 126–135.
[9] Miner, E. D., 1990, *Uranus: The Planet, Rings and Satellites* (2nd Edition), Wiley-Praxis Series in Astronomy and Astrophysics, p. 57.
[10] Elliot, J. L., 1982, "Rings of Uranus: A review of occultation events", in *Uranus and the Outer Planets*, edited by G. Hunt, Cambridge University Press, pp. 237–256.
[11] Elliot, J. L., Nicholson, P. D., 1984, "The rings of Uranus" in *Planetary Rings*, edited by Greenberg and Brahic, The University of Arizona Press, Tucson, pp. 25–72.
[12] French, R. G., Elliot, J. L., Levine, S. E., 1986, "Structure of the uranian rings", *Icarus* **67**, 134–163.
[13] French, R. G., Elliot, J. L., French, L. M., Kangas, J. A., Meech, K. J., Ressler, M. E., Buie, M. W., Frogel, J. A., Holberg, J. B., Fuensalida, J. J., Joy, M., 1988, "Uranian ring orbits from Earth-based and Voyager occultation observations", *Icarus* **73**, 349–378.
[14] Goldreich, P., Tremaine, S., 1979, "Towards a theory for the uranian rings", *Nature* **277**, 97–99.
[15] Miner, E. D., 1990, *Uranus: The Planet, Rings and Satellites* (2nd Edition), Wiley-Praxis Series in Astronomy and Astrophysics, p. 256.

[16] *Ibid.*
[17] *Ibid.*
[18] *http://pds-rings.seti.org/uranus/uranus_tables.html*
[19] *Ibid.*
[20] UC Berkeley News Press Release, December 22, 2005, "Keck Telescope captures faint new ring around Uranus".
[21] *Ibid.*

## 3.6 BIBLIOGRAPHY

Elliot, J., Kerr, R., 1984, *Rings—Discoveries from Galileo to Voyager*, The Massachusetts Institute of Technology Press.

Greenberg, R., Brahic, A., (eds.), 1984, *Planetary Rings*, University of Arizona Press, 784 pp.

Miner, E. D., 1990, *Uranus: The Planet, Rings and Satellites* (2nd Edition), Wiley-Praxis Series in Astronomy and Astrophysics, 360 pp.

## 3.7 PICTURES AND DIAGRAMS

Figure 3.1 *http://www.nasm.si.edu/research/ceps/etp/tools/img/KAOflight.jpg*
Figure 3.2 *http://www.nasm.si.edu/research/ceps/etp/discovery/discimg/EP0003.jpg*
Figure 3.3 Miner, E. D., 1990, *Uranus: The Planet, Rings and Satellites* (2nd Edition), Wiley-Praxis Series in Astronomy and Astrophysics, fig. 4.3.
Figure 3.4 Miner, E. D., 1990, *Uranus: the Planet, Rings and Satellites* (2nd Edition), Wiley-Praxis Series in Astronomy and Astrophysics, fig. 4.4.
Figure 3.5 *http://photojournal.jpl.nasa.gov/jpeg/PIA01985.jpg*
Figure 3.6 *http://photojournal.jpl.nasa.gov/jpeg/PIA00142.jpg*
Figure 3.7 *http://imgsrc.hubblesite.org/hu/db/2005/33/images/b/formats/full_jpg.jpg*
Figure 3.8 *http://www.nasa.gov/mission_pages/hubble/Uranus_ring.html*
Figure 3.9 *http://www.berkeley.edu/news/media/releases/2005/12/22_rings.shtml*

# 4

# The discovery of the Jupiter ring system

## 4.1 FIRST OBSERVATIONS OF JUPITER'S RING

Everyone knew that only Saturn (and, more recently, Uranus) had rings. Even models used to predict long-term stability of planetary rings indicated that the Jovian system could not maintain one [1]. Nonetheless, Tobias Owen, a co-investigator on the Voyager Imaging Science Subsystem (ISS) Team, thought that a planet the size of Jupiter must have debris around it [2]. He postulated that Jupiter might have material about it in the equivalent region of Saturn's B ring. Although most investigators thought that Jupiter did not have such material, Brad Smith, the ISS Principal Investigator, had his own reason to support such an observation. In retrospect, Dr. Smith stated that a search for a Jovian ring was not made "with any great expectation of a positive result but more for the purpose of providing a degree of completeness to Voyager's survey of the entire Jupiter system" [3].

The "B-ring" distance from Jupiter was a logical place to search for a ring since it was inside the planet's Roche limit. Inside the Roche limit, the difference in gravitational forces on the front and back of a satellite is so extreme that a satellite drifting interior to this limit will be torn apart. In Saturn's case, the main ring system, including the F ring, lies inside this limit. Of course, the limit would be different for each planet. It also would be dependent on a satellite's size and composition. A Jupiter ring observation would have to be designed to span a region that would account for every possible size and composition of the original satellite.

Once designed, the ring observation had to be inserted somewhere in a Voyager 1 timeline of events already crowded with higher priority observations. Since a Jupiter ring had never been seen from Earth, a ring, if it existed, would be optically thin. Consequently, the best time to try to detect such a ring would be during ring-plane crossing. During that time period, the observation would look through as much as possible of the ring material that orbited near the planet's equatorial plane. To further

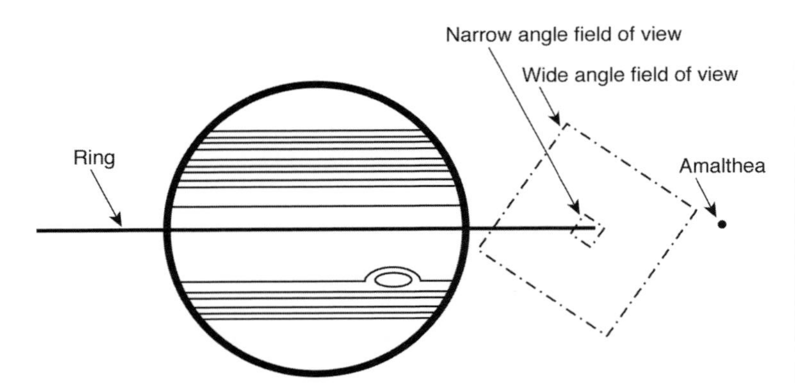


**Figure 4.1.** This schematic drawing shows the approximate size and orientation of the ISS wide-angle and narrow-angle fields of view on the Jovian system.

increase the chance of success, the observation should incorporate a long imaging exposure.

With numerous competing science observations and an allocation of only two ring-plane crossing events, Tobias Owen and Ed Danielson, another ISS co-investigator, designed a ring observation that would fit into the first ring-plane crossing event. They initially requested a four-image observation but competing science demands resulted in reducing the observation time down to just one exposure [4]. Fortunately, the investigators knew they could recover at least one of the lost imaging frames by simultaneously shuttering both the wide-angle and narrow-angle cameras. The cameras' fields of view were oriented diagonally (to span the region of the proposed ring) and centered between Jupiter and its innermost satellite, Amalthea [5] (see Figure 4.1). The exposure times for both cameras were set to 11.2 minutes to maximize their chance of detecting faint ring material that might exist [6].

The Voyager 1 encounter began sixty days before its March 5, 1979, 4:42 a.m. PST rendezvous with Jupiter [7]. The start date had been chosen to coincide with the time when Voyager 1's imaging capabilities exceeded those of ground-based observatories. Unfortunately, this would have placed the start of the encounter just before the Christmas holidays. In order to give the operations teams one last break before the historic and exceedingly hectic Jupiter encounter, the start time was delayed to early January 1979. The single-ring search observation was known as JRINGS and was scheduled to execute at 16 hours and 52 minutes before Jupiter closest approach [8].

The Jupiter encounter was divided into four phases: Observatory, Far Encounter, Near Encounter and Post Encounter. The Near Encounter Phase contained the closest approach point and was the most scientifically valuable part of the encounter. As Voyager approached its rendezvous with Jupiter, command sequences in its onboard computer executed one by one. At the designated time in the Near Encounter Phase, Voyager faithfully shuttered its cameras for the JRINGS observation. The wide-angle image was saturated, but the narrow-angle image found a ring [9] (shown in Figure 4.2).

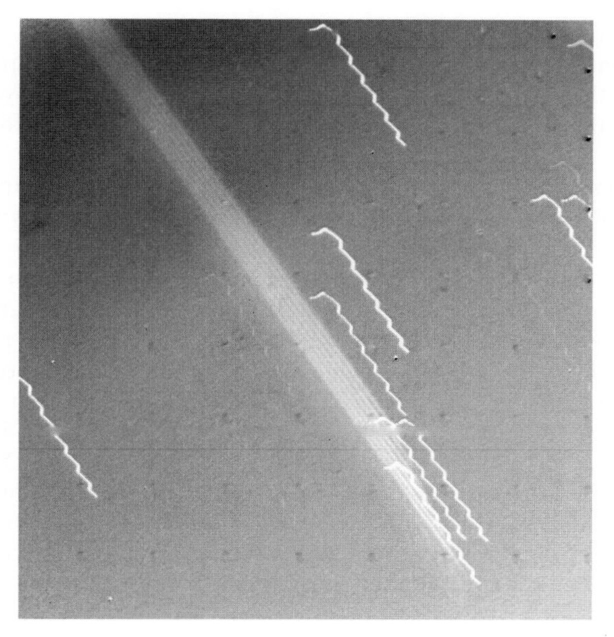

**Figure 4.2.** Discovery image of the ring of Jupiter. The ring is the broad white band that crosses the center of this narrow-angle frame. The long duration nature of the image combined with spacecraft motion to make the star streaks appear jagged and create multiple images of the thin ring. (PIA02251)

In the discovery image the ring is approximately 1,212,000 km from the spacecraft and 57,000 km from Jupiter's visible cloud tops [10]. This image also showed that the ring particles weakly back-scatter light; a later Voyager 2 image (Section 4.2) displayed strong forward-scattering, which in turn meant that the ring particles were extremely small, perhaps on the order of micrometer-sized dust [11].

The Jovian ring discovery was announced on March 7, 1979, and appeared in the next day's newspaper. Two scientists who were observing at the Mauna Kea Observatory on Hawaii decided to see if they could corroborate Voyager 1's claim. In just two days, Dr. Eric Becklin and Dr. Gareth Wynn-Williams—using an infrared detector sensitive at 2.2 μm—confirmed the historic ring discovery [12].

## 4.2 THE SECOND ENCOUNTER WITH JUPITER'S RING

The exciting discovery of a Jupiter ring, combined with a number of other important scientific discoveries, led to a strong desire to redo the previously designed Voyager 2 encounter sequences. The project had four months to understand the first Jupiter encounter results, figure out what observations should be added to answer questions raised by that encounter; and then make the painful decisions about what observations to delete to make room for the new ones. As an indication of how low the expectations were of finding a Jupiter ring, Voyager team members had planned no ring observations for the second Jupiter encounter [13]. The large quantity of high-priority science that could only be captured by Voyager 2, even if Voyager 1 collected all of its planned encounter observations, had led investigators to forgo future ring observations.

The Voyager 2 encounter with Jupiter began on April 24, 1979, approximately 76 days before closest approach [14]. The dimness of the rings in the Voyager 1 image led scientists to speculate that the ring particles were very small, there was therefore a desire to image the rings from beyond Jupiter, where expected forward-scattering would make the rings brighter. Ring observations would have to wait until after the July 9, 1979, closest approach when the spacecraft was behind the planet.

On July 10, 1979, the Voyager 2 cameras were commanded to look back towards the Sun (which was hidden from the spacecraft by Jupiter) and search for aurorae and lightning. Nestled into these high-phase-angle observations were imaging frames designed to detect scattered light from the newly discovered ring. Data transmitted back to Earth showed that these high-phase-angle ring observations were a spectacular success. A series of narrow-angle frames revealed a faint ring structure that was all but absent from the Voyager 1 images. Figure 4.3 shows a narrow-angle frame that captures one edge of the Jupiter ring system.

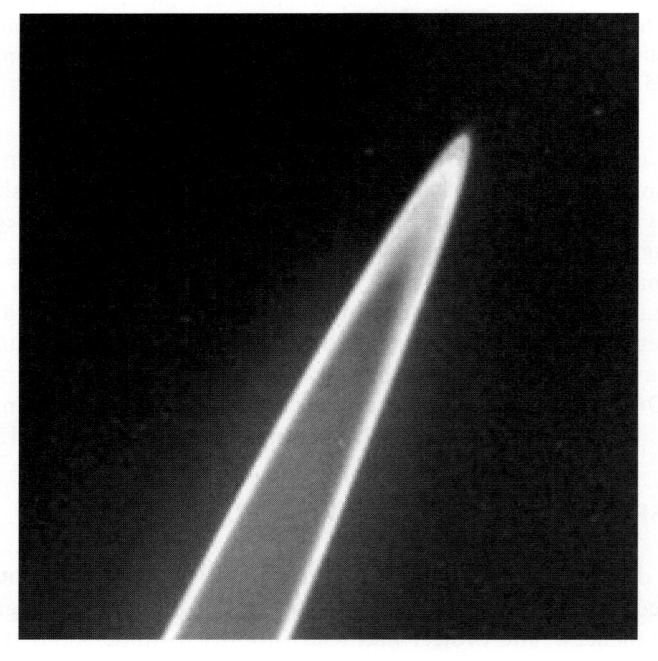

**Figure 4.3.** This Voyager 2 image of one of Jupiter's ring ansae was shuttered when the spacecraft was 1.5 million km away. The ring has a sharp outer edge with a much more diffuse inner edge. (PIA00377)

From this and other narrow-angle frames, imaging scientists were able to determine that the Jupiter ring was composed of two very tenuous rings. The inner ring extended from the upper atmosphere to a distance of 53,000 km above the cloud tops [15]. The outer ring, which is the more pronounced portion of the ring system, extended further by another 5,000 km. The thickness of the outer ring was not determined but was probably less than 30 km and more likely less than 1 km [16]. There were also some indications that the ring particles extended out of the equatorial plane [17].

Subsequent analysis of Voyager 2 data revealed a third ring residing outside the two already discovered. This additional ring was extremely faint and extended to approximately 210,000 km [18]. Barely observable, this "gossamer" ring would require more data from a more sensitive instrument to ascertain its true nature.

The two Voyager Jupiter encounters were a dramatic success. They greatly increased the understanding of the planet's atmosphere, magnetosphere, satellites, and rings. However, every question that was answered seemed to give rise to many more questions. Of the new questions raised, many were about the rings themselves. These questions included, "What was the source of the ring material?"; "Why were there multiple rings?"; and "Why were some ring particles out of the equatorial plane?". Another mission to Jupiter was needed to attempt to answer these questions, a mission that would spend years, not months, making detailed observations of all aspects of the system, including its rings.

## 4.3 GALILEO ENCOUNTER

The Galileo spacecraft was the first vehicle ever designed to orbit an outer planet. Its prime mission was to release an atmospheric probe into Jupiter's atmosphere and then spend the next two years in orbit about the planet trying to answers as many questions as possible that were raised by Voyager's Jupiter encounters. Galileo science objectives were divided into three main disciplines: atmosphere, magnetosphere, and satellites. The satellite discipline was also responsible for determining the origin, composition, and dynamics of the Jovian rings.

Unfortunately, prior to its Jupiter arrival the Galileo spacecraft was unable to unfurl its high-gain antenna (HGA). Galileo would have to use its rigid low-gain antenna (LGA). The LGA was primarily designed for spacecraft emergencies when the spacecraft might not be able to locate Earth. This small antenna could transmit data only at low rates but could radiate its signal over a much wider area. If it was needed, all science transmissions would be terminated and only engineering data would be sent to Earth. The loss of the HGA meant that the spacecraft could collect vastly more data than it could possibly transmit to Earth. Data rates had to be reduced from the planned 134,000 bits/s with the HGA down to approximately 30 bits/s with the much lower capability of the LGA [19].

Drastic times called for drastic measures. Torrence Johnson, the Galileo Project Scientist, allocated to each science discipline an exceedingly small number of bits. The total number of bits was dictated by the amount of data that could be collected and transmitted to Earth between successive Jupiter orbits. Each science discipline had to determine their highest-value observations on each particular orbit. Then an appropriate fraction of the allocated bits was assigned to each instrument to capture its relevant science data. Investigators' decisions were based on the discipline's highest-priority science, the particular satellite the spacecraft was going to encounter on that orbit, and the desired instrument viewing geometry. Needless to say, the Galileo investigators agonized over what data to collect and what to abandon.

The Galileo spacecraft arrived at Jupiter on December 7, 1995. Its prime orbital mission was to last for two years and included 11 orbits about the planet. In reality, the spacecraft lasted much more than the planned two years. Galileo had three mission extensions. The first was called the Galileo Europa mission (GEM) and was designed to focus orbits 12 through 25 on determining if Europa had subsurface liquid water [20]. The next extension was called the Galileo Millennium mission (GMM) and added four more science data collection orbits about the planet [21]. The final extension, known as the Galileo Millennium mission extension, added five more orbits [22].

## 4.4 GALILEO RING OBSERVATIONS

During Galileo's many years of operations, its science instruments revealed many additional details about the three distinct parts of the planet's ring system. The inner donut-shaped ring, called the Halo ring, may extend all the way to the planet's cloud tops. Outside of the Halo ring is an optically thicker and flatter ring, known as the Main ring. The third and most distant part of the system is the tenuous Gossamer ring. The Gossamer ring in actuality is composed of two overlapping rings. One begins at the inner edge of the Main ring and extends outward to the orbit of Amalthea. Its physical thickness is greater than that of the Main ring and thus engulfs the Main ring. The other Gossamer ring also begins at the inner edge of the Main ring but extends out past Amalthea to the orbit of Thebe. This outer Gossamer ring is also thicker than the inner Gossamer ring and thus contains both the inner Gossamer ring and the Main ring [23] (see Figure 4.4).

Galileo spacecraft observations indicate that the rings are short lived and are composed of small dust particles. Particle size was inferred from the light-scattering properties of the individual rings. It is believed that the small ring particles are liberated from the four inner moons (Metis, Adrastea, Amalthea and Thebe), when meteoroids collide with their surfaces [24].

Figure 4.5 is a mosaic of five images and is displayed twice. The top set of images contains the raw imaging data while the bottom is the same data annotated with the now-accepted names of three of the four rings and the four innermost moons. The exposures of the individual frames were successively increased to show the faint material in orbit about the planet. The vertical arc on the left is from sunlight transmitted through Jupiter's atmosphere. The first frame, starting from the left, has the typical image sensitivity and shows the Main ring. The next image has ten times the sensitivity and shows an over-exposed Main ring with fainter material further away from the planet. The outer edge of the faint Halo ring is also visible.

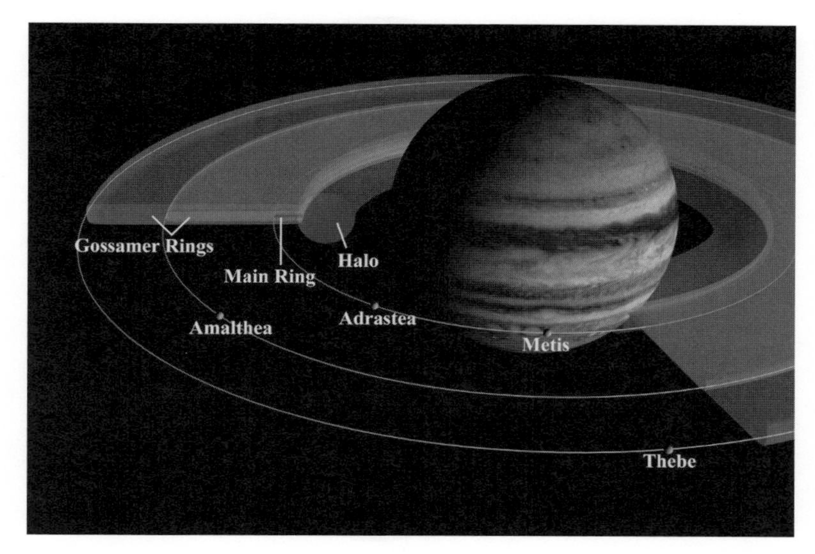


**Figure 4.4.** This schematic cut-away view shows the components of Jupiter's ring system. Debris from meteoroid bombardment of the four inner satellites, Metis, Adrastea, Amalthea and Thebe, probably supplies the ring system with material. (PIA01627)

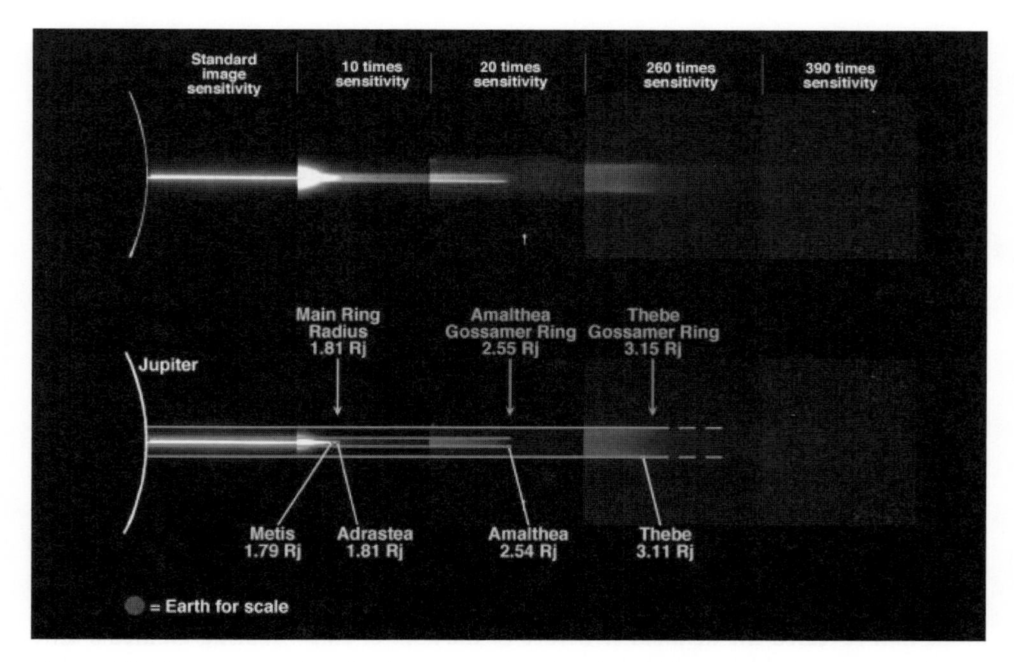


**Figure 4.5.** The top panel is a five-frame mosaic taken by Galileo on October 5, 1996, during orbit C3 (i.e., third orbit, Calisto encounter). The bottom image contains the location of the four small moons that are responsible for producing material for the rings. (PIA01623)

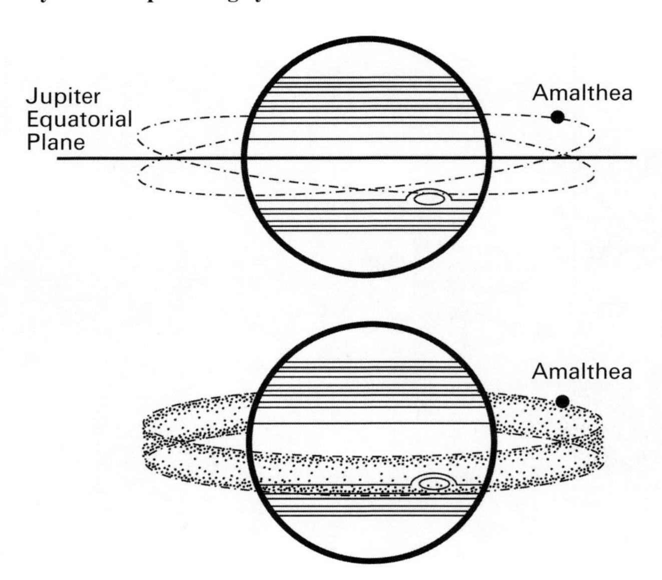


**Figure 4.6.** The extended thicknesses of the Gossamer rings are dictated by the inclination of Amalthea and Thebe. This diagram illustrates how Amalthea produces the edge of the inner Gossamer ring. (PIA01628)

This is followed by three more frames which provide a 20 times, 260 times, and a 390 times more sensitive exposure.

The image shows that the inner Gossamer ring abruptly ends at the orbit of Amalthea while the outer Gossamer ring ends at the orbit of Thebe. This, coupled with the fact that Gossamer rings have a thickness corresponding to the entire northern and southern extent of the orbits of Amalthea and Thebe, respectively, revealed an unusual mechanism for their structure. Micro-meteoroids impacting the surfaces of the two satellites eject material from the moons' surfaces. Since the orbits of these satellites are inclined to the planet's equatorial plane, so are the orbits of the liberated dust particles [25]. Since material is ejected from both satellites from all parts of their orbit, the continuous loss of particles produces a ring structure that is much thicker than for the Main ring (see Figure 4.6). The two Gossamer rings do not have a flat elliptical or circular shape like Jupiter's Main ring or the rings in Saturn's ring system, but are more like flat-edged disks [26]. The Main ring has no such dispersion: both Metis and Adrastea are in Jupiter's equatorial plane [27].

During its 35 orbits around Jupiter, the Galileo spacecraft passed through Jupiter's shadow only three times (orbit 3, 4, and 5) [28]. These shadow passages provided the sole opportunities for the sensitive science instruments to look close to the direction of the Sun without being damaged. It is these high-phase-angle observations that revealed the detailed structure of the Jupiter rings. During Galileo's third orbit of Jupiter, which included a close encounter of Callisto, the spacecraft looked back towards the Sun during shadow passage and imaged Jupiter and its entire ring system [29] (see Figure 4.7).

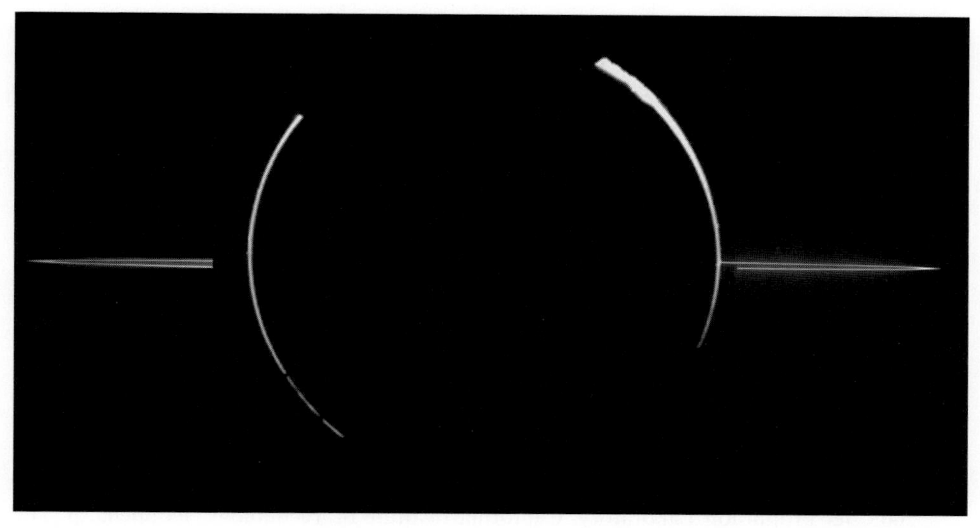

**Figure 4.7.** This Jupiter mosaic taken on November 9, 1996, consists of two Galileo spacecraft images taken through the clear filter when the spacecraft was behind the planet as seen from the Sun. The Main ring system can be seen surrounding the planet. (PIA01621)

The Galileo Millennium Mission Extension included the controlled demise of the spacecraft. After eight years of orbital operations, the spacecraft was running out of attitude control gas. The loss of propellant would result in the loss of control of the spacecraft and the eventual collision with one of the Jovian satellites. Scientists believed that there would be a greater consequence of contaminating one of Jupiter's moons (and their possibility of life) than of having the craft burn up in the giant planet's atmosphere. Therefore, after saying goodbye, ground controllers commanded the spacecraft to plunge into the planet's atmosphere. The final orbit targeted Galileo to impact Jupiter on September 21, 2003, at 12:49:37 PDT [30].

## 4.5 NOTES AND REFERENCES

[1] Morrison, D., Samz, J., *Voyage to Jupiter*, NASA SP-439, 1980, p. 84.
[2] Personal communication with Tobias Owen.
[3] Elliot, J., Kerr, R., 1984, *Rings—Discoveries from Galileo to Voyager*, The Massachusetts Institute of Technology Press, 1984, p. 100.
[4] *Ibid.*
[5] Personnel communication with Candice Hansen-Koharcheck.
[6] Morrison, D., Samz, J., (1980) Voyage to Jupiter, NASA SP-439, p 85.
[7] *Ibid.*, p. 63.
[8] *Ibid.*, p. 85.
[9] Personnel communication with Charles C. Avis.
[10] Caption found at *http://photojournal.jpl.nasa.gov/catalog/PIA02251*

[11] *Flight Science Office Science and Mission Systems Handbook—Voyager Uranus/Interstellar Mission*, Voyager Document PD 618-128 (JPL D-498), Revision C, dated 1985 July 1, Jet Propulsion Laboratory, California Institute of Technology, Pasadena, p. 2.3.

[12] Morrison, D., Samz, J., 1980, *Voyage to Jupiter*, NASA SP-439, p. 86.

[13] Elliot, J., Kerr, R., 1984, *Rings—Discoveries from Galileo to Voyager*, Massachusetts Institute of Technology Press, p. 104.

[14] Morrison, D., Samz, J., 1980, *Voyage to Jupiter*, NASA SP-439, p. 93.

[15] *Ibid.*, p. 136.

[16] *Ibid.*

[17] *Flight Science Office Science and Mission Systems Handbook—Voyager Uranus/Interstellar Mission*, Voyager Document PD 618-128 (JPL D-498), Revision C, dated 1985 July 1, Jet Propulsion Laboratory, California Institute of Technology, Pasadena, p. 2.3.

[18] Showalter, M. R., Burns, J. A., Cuzzi, J. N., Pollack, J. B., 1985, "Discovery of Jupiter's 'gossamer' ring", *Nature* **316**, 526–528.

[19] Personal communication with Brian Paczkowski.

[20 *Galileo Project Final Report*, Galileo Document JPL D-28516, Volume 1, 21 September 2003, Jet Propulsion Laboratory, California Institute of Technology, Pasadena, p. 115.

[21] *Ibid.*, p. 153.

[22] *Ibid.*, p. 174.

[23] Caption found at *http://photojournal.jpl.nasa.gov/catalog/PIA03001*

[24] A brief description of the Jupiter ring system can be found at URL *http://solarsystem.nasa.gov/planets/profile.cfm?Object = Jupiter&Display = Rings*

[25] Caption found at *http://photojournal.jpl.nasa.gov/catalog/PIA01628*

[26] *Ibid.*

[27] Cornell News, *Galileo finds Jupiter's rings formed by dust blasted off small moons, http://www.news.cornell.edu/releases/sept98/jupiter_rings.html*

[28] *Galileo Project Final Report*, Galileo Document JPL D-28516, Volume 1, 21 September 2003, Jet Propulsion Laboratory, California Institute of Technology, Pasadena, Appendix A-5.

[29] *Ibid.*

[30] *Ibid.*, p. 203.

## 4.6 BIBLIOGRAPHY

Elliot, J., Kerr, R., 1984, *Rings—Discoveries from Galileo to Voyager*, The Massachusetts Institute of Technology Press, 1984.

*Flight Science Office Science and Mission Systems Handbook—Voyager Uranus/Interstellar Mission*, Voyager Document PD 618-128 (JPL D-498), Revision C, dated 1985 July 1, Jet Propulsion Laboratory, California Institute of Technology, Pasadena.

Guiliano, J. A. (compiled), 1978, *Voyager 1 and 2—Jupiter Science Link Dictionary Index*, JPL Document 618-805.

Morrison, D., Samz, J., 1980, *Voyage to Jupiter*, NASA SP-439.

## 4.7 PICTURES AND DIAGRAMS

Figure 4.1 [6] Morrison, D., Samz, J., 1980, *Voyage to Jupiter*, NASA SP-439, p. 84. (modified by Robbii R. Wessen).

Figure 4.2 *http://photojournal.jpl.nasa.gov/jpeg/PIA02251.jpg*

Figure 4.3 *http://photojournal.jpl.nasa.gov/jpeg/PIA00377.jpg*

Figure 4.4 *http://photojournal.jpl.nasa.gov/jpeg/PIA01627.jpg*

Figure 4.5 *http://photojournal.jpl.nasa.gov/jpeg/PIA01623.jpg*

Figure 4.6 *http://photojournal.jpl.nasa.gov/figures/satinc.html* (modified by Robbii R. Wessen).

Figure 4.7 *http://photojournal.jpl.nasa.gov/jpeg/PIA01621.jpg*

# 5

# The discovery of the Neptune ring system

## 5.1 FIRST OBSERVATIONS OF NEPTUNE'S RING

Neptune was an enigma. Jupiter, Saturn, and Uranus all had rings. If there was justice in the universe, Neptune should also have rings. But if it didn't, some explanation would have to be developed to account for Neptune's inadequacy. In either case, astronomers first had to determine if there were rings. Unfortunately, the search for these rings was made more difficult by the planet's extreme distance from the Sun. The only viable approach for discovering rings was by stellar ring occultation observations. This technique requires the planet to cross in front of a stable star. If there are rings, the star will appear to flicker prior to the occultation and then again after the star emerges from behind the planet's disk. A stable star of known output is essential if the data are to be interpreted correctly. One could easily mistake brightness variations of a star for ring occultation events.

At first blush, searching for rings with stellar ring occultations would seem to have an equally high probability of success at Neptune as it had at Uranus. After all, it was this technique that was used to discover Uranus's rings. However, there are two major geometric differences between Uranus and Neptune that make this technique for discovering rings at Neptune far more difficult. First, Neptune is one-and-one-half times as far away from the Sun (and Earth) as Uranus is. This makes Neptune's disk appear smaller, which in turn means that fewer star occultations occur. As a matter of fact, stellar ring occultions at Neptune are so rare that they occur on average only a few times a year.

The other factor is tied to the planets' axial tilts. Uranus's axial tilt is 97.9° to its orbital plane [1]. This makes the ring system appear open as seen from Earth, at least away from the times that Uranus's rings are edge-on to Earth (1965, 2007, 2049, etc.). Because of this, Uranus's rings generally subtend a relatively larger amount of sky than Neptune's; this factor increases the odds that Uranus's rings will occult a star. For Neptune the situation is not as favorable. If Neptune had a ring, it would most likely be equatorial. In that case, a ring system would appear more nearly edge-on to Earth-based observers since Neptune's axial tilt is only 29.6° [2]. Thus, the Uranian ring system subtends four times the amount of sky as compared with a proposed Neptune ring system (about 8 arc-seconds for Uranus as opposed to only 2 arc-seconds for Neptune) [3].

Astronomers started to look for Neptune rings immediately following the planet's discovery in 1846. As a matter of fact, two weeks after the planet's discovery, astronomer William Lassell (1799–1880) reported finding such a ring [4]. However, there were no corroborating data to support his claim. This was not the first time an illusive ring was thought to be encircling a newly discovered planet. William Herschel (1738–1822) thought he saw a ring around his newly discovered "Georgian Star". Though his telescope was strong enough to discover Uranus it wasn't strong enough to discern a ring. Richard Baum, Vice President of the British Astronomical Association from 1993 to 1995, thought Herschel's ring was more likely astigmatism associated with his telescope rather then an actual physical ring. This type of defect may also have led Lassell to make his claim about Neptune.

The earliest documented observation of a possible Neptune ring occurred more than 120 years later in 1968. In that year, astronomers at Villanova University were trying to obtain a better measurement for Neptune's diameter. Their approach was to accurately measure the duration of a Neptune occultation of a 7th magnitude star. The observation was a success, but no one noticed a 30% decrease in stellar brightness for 2 minutes and 48 seconds, beginning three minutes after the planet occultation.

Thirteen years later on May 10, 1981, another Neptune occultation was observed photometrically. This time the brightness variation prior to the occultation was noted, albeit well after the event. Villanova team members, excited by this finding, went back to the data archives for data from the 1968 event. Sure enough, something had occulted the star prior to the occultation of Neptune's atmosphere. Astronomers immediately went to work to determine the geometry of a possible ring. Results indicated that if Neptune had an equatorial ring it would be located about 29,800 km to 36,125 km from the planet's center [5]. Of course, this was just an educated guess. The dimming could have been from variability of the occulted star, an experimental error, or possibly something else.

## 5.2 MORE STELLAR RING OCCULTATIONS

Luck was on the side of the Villanova team. Another Neptune stellar occultation was to occur only two weeks later (on May 24, 1981). It would only be a grazing event, but the star was very stable and well-behaved. The team, led by Harold J. Reitsema, could potentially confirm the Villanova team's discovery. Once the observations were made, measurements made by Reitsema's team did show a brief drop in brightness, but the data didn't look like what one would expect for a ring occultation. The interpretation was more consistent with the unlikely discovery of a new Neptune satellite, one with a diameter of approximately 180 km [6]. The only way to substantiate this discovery was with additional measurements. Unfortunately, the small size of the satellite, the poor knowledge of its orbit, and the paucity of occultation events made confirmation unlikely. The Villanova team would have to wait eight more years for the Voyager 2 Neptune encounter to have a chance at confirmation of their potential discovery.

Meanwhile, the two stellar occultations and their atypical results served to increase interest and speculation about the cause of these brightness decreases. The next observed stellar occultation event was on June 15, 1983, and W. B. Hubbard was ready. Hubbard and his colleagues observed the event from six different locations. The multiple viewing angles allowed Hubbard to search as close as 0.03 radii of the planet's surface for rings. Unfortunately, no indications of rings were found [7]. Over the next six years many observations were made. Table 5.1 lists these attempts and their varied results. These inconsistencies vividly demonstrated how difficult these measurements are and how difficult it is to determine from Earth whether or not Neptune had rings.

With only a third of the occultation events showing possible brightness variations, the ring system, if it existed, was not typical. It was the job of the theoreticians to combine all the observed data and piece together a model that could explain these observations. However, no single model would fit all the findings. If Neptune's rings existed, the truth about them would contain some combination of the following hypotheses:

(1) Two shepherding satellites, each between 100 to 200 km in diameter could confine the ring (they would be too small to be seen from Earth) [9].
(2) The ring system was either incomplete or at least highly azimuthally variable [10].
(3) Newly discovered satellites might produce some of the observations.
(4) Incomplete rings could consist of a series of short arcs having a width of 100 km that center on co-rotation resonances of a single satellite in an inclined orbit [11].

Ring scientists were clearly puzzled. One of them, Philip D. Nicholson, stated that no more than three of all the stellar occultation events were probably real. However, based on this meager data set, he postulated that if Neptune did have rings it would have at least three distinct ring arcs, the outer-most of which was approximately 72,500 km from the center of the planet. The other two could not be resolved into a single location. In 1984 the situation was summed up by Elliot and Kerr who said, "If Neptune has rings, they almost certainly will not be discovered from the ground" [12].

## 5.3 THE VOYAGER 2 NEPTUNE ENCOUNTER

Voyager 2 was going to be the first spacecraft from Earth to visit Neptune. However, a Voyager 2 Neptune encounter was not always in the approved plan or in the budget. Fortunately, mission planners continued to keep the possibility alive from the mission's very inception back in the early 1970s. Almost 20 years later very little had changed to increase our understanding of the planet. Voyager 2 was going to pass

**Table 5.1.** Neptune stellar occultations observed from 1968 to 1989 [8].

| Date of occultation | Event seen | Corresponds to a "ring" location at (km) | Observed from |
|---|---|---|---|
| 1968 Apr 7 | Possible occultation | 29,800–36,125 | MJO and Japan (2) |
| 1980 Aug 21 | Possible occultation | | MSO |
| 1981 May 10 | — | N/A | AAO, IRTF, MSO, UH |
| 1981 May 24 | — | N/A | CTIO |
| 1981 May 24 | Discovery of a satellite | 73,500 | CAT, ML |
| 1983 Jun 15 | — | N/A | AAO, CFH, IRTF, KAO, UH, Australia (2) |
| 1983 Sep 12 | Possible occultation | 28,700–31,700 | UPSO |
| 1984 Apr 18 | Possible occultation | 55,200 | PAL |
| 1984 May 11 | — | N/A | PAL |
| 1984 Jul 22 | Confirmed event | 67,000–75,000 | CTIO, ESO |
| 1985 Jun 7 | Possible occultation | 62,600 or 75,000 | SAAO |
| 1985 Jun 25 | — | N/A | SAAO |
| 1985 Jul 30 | — | N/A | PAL, IRTF |
| 1985 Aug 10 | — | N/A | IRTF |
| 1985 Aug 20 | Confirmed event | 62,900–63,000 | CFH, CTIO, ESO, IRTF, LOW, MWO |
| 1986 Apr 23 | — | N/A | CFH, IRTF, UH |
| 1986 May 4 | — | N/A | KPNO, ESO |
| 1986 Jul 27 | Confirmed event | 42,300 | IRTF, PAL, UH |
| 1986 Aug 23 | — | N/A | IRTF, UH, UKIRT |
| 1987 May 23 | — | N/A | IRTF, PAL |
| 1987 Jun 22 | Possible Triton occ. | 245,800 | ESO |
| 1987 Jul 9 | Confirmed event | 35,200 | KPNO, MMT, PAL, UKIRT |
| 1987 Aug 29 | Possible occultation | | IRTF (?) |
| 1988 May 26 | — | N/A | IRTF (?) |
| 1988 Jul 9 | Possible occultation | 63,170 or 63,120 | MMT, PAL |
| 1988 Aug 2 | — | N/A | ESO |
| 1988 Aug 25 | — | N/A | CFH, IRTF |
| 1988 Sep 12 | — | N/A | OHP, OPMT |
| 1989 Jul 8 | — | N/A | ESO, OPMT |

AAO Anglo-Australian Observatory
CAT Catalina Station
CFH Canada–France–Hawaii Telescope
CTIO Cerro Tololo Interamerican Observatory
ESO European Southern Observatory
IRTF Infrared Telescope Facility
KAO Kuiper Airborne Observatory
KPNO Kitt Peak National Observatory
LOW Lowell Observatory
MJO Mount John Observatory
ML Mount Lemmon
MMT Multiple Mirror Telescope
MSO Mount Stromlo Observatory
MWO Mount Wilson Observatory
OHP Observatoire de Haute Provence
OPMT Observatoire du Pic du Midi
PAL Palomar Observatory
SAAO South Africa Astronomical Observatory
UH University of Hawaii 88-inch
UKIRT United Kingdom Infrared Telescope
UPSO Uttar Pradesh State Observatory

Neptune regardless of how much its Earth-bound controllers knew about it. With this in mind, the project did the best it could and began planning the encounter sequences for the upcoming rendezvous.

The project first evaluated the condition of the spacecraft and the available ground assets. On the science side of the house, the Voyager science investigators held their own Neptune Science Working Group meetings between August 1986 and July 1987. At these meetings scientists discussed what was known about Neptune from the sparse and often conflicting ground-based observations; the type of capabilities the spacecraft had; and what were the highest value observations the craft could perform to reveal the greatest amount of information about Neptune. The Voyager 2 science goals for Neptune ring observations were to obtain information on the following (if rings existed) [13]:

(1) Ring-arc structure: both radial and azimuthal profiles and their changes with time.
(2) Ring satellite search for moonlets in or near the rings.
(3) Search for additional ring material.
(4) Orbital kinematics—improve orbit models, compare ring motions with satellite motions.
(5) Ring particle motions—importance of self-gravity, vertical thickness of rings, collisions between particles.
(6) Ring particle properties: size, shape, reflectivity, temperature, and composition.
(7) Ring environment: interaction with magnetospheric plasma or with an extended neutral atmosphere.

By the time Voyager reached Neptune, the spacecraft had been in space for almost 12 years; the light levels were down by a factor of 900 as compared with Earth; and data rates had fallen from 115,000 bits/s at Jupiter to about 3,200 bits/s at Neptune due to the greater distance [14]. All of these factors severely limited the amount and quality of data that could be collected and was clearly not acceptable. So, engineers at the Jet Propulsion Laboratory began to modify the flight software to make Voyager 2 a much more capable craft.

Spacecraft engineers knew that there were only a limited number of things that could be done to the spacecraft to compensate for the more difficult observing conditions. The lower light levels could be overcome with longer exposures. But longer exposures produce greater smear. To compensate for this, engineers programmed the spacecraft to gently rotate as the images were shuttered. This image motion compensation (IMC) was analogous to a human photographer panning her camera to remove blur out of the subject while the background is excessively smeared. The major liability of this technique is that since the spacecraft was turning, the spacecraft's high-gain antenna would drift away from Earth-point. The lack of Earth-point meant that all data collected during an IMC observation had to be placed on the spacecraft's digital tape recorder (DTR). This technique had been successfully executed at Uranus and was planned for use again at Neptune.

However, the greater spacecraft range from Earth resulted in slower data rates, which in turn required a longer time to return the data from the spacecraft's recorder. This turned the data storage on the tape recorder into a valuable resource. Knowing this, engineers designed a new type of IMC to "nod" back and forth. This nodding motion allowed the angular rates to be taken out of the target without drifting too far away from the Earth. Between each "nod", the spacecraft would be commanded to reposition for the next imaging frame. Nodding image motion compensation (NIMC) could be done without interrupting the downlink telemetry and saved the DTR for other more valuable observations.

Engineers also developed another type of IMC that did not involve moving the spacecraft at all. The idea was to command the spacecraft to move only its scan platform at its slowest possible rate. This did produce a distinctive "jerkiness" in the images but maneuverless image motion compensation (MIMC) was reserved for those observations of rapidly moving targets that otherwise would be useless. The moving scan platform had the imaging cameras (ISS), the infrared interferometer and spectrometer (IRIS), the photopolarimeter (PPS), and the ultraviolet spectrometer (UVS) mounted on it, but IMC would be primarily used to help the imaging instruments. Table 5.2 shows the Voyager 2 Neptune ring observations that were designed with IMC. In this table, retargetable observations were sequences on the spacecraft with parameters that could be updated with better pointing information. Retargeting ensured successful pointing of the instruments at the target body. The start times are given in encounter relative time which is the time relative to the spacecraft's arrival at its closest approach point.

**Table 5.2.** Ring observations that were designed with image motion compensation [15].

| Observation | Observation name | Start time (enc. rel.) | Type of IMC |
|---|---|---|---|
| Ring retargetable, ISS | VRRET1 | −07:16.8 | NIMC |
| Ring retargetable, ISS | VRRET1 | −06:40.8 | NIMC |
| Ring retargetable, ISS | VRRET1 | −03:12.8 | NIMC |
| High-phase ring observation, ISS | VRHIPHAS | +00:40.0 | MIMC |
| Ring plane crossing, ISS | VRING2 | +01:15.2 | MIMC |

Mission planners also knew that there were things that could be done that did not involve the spacecraft. One of these was to make all three of the large 64-m Deep Space Network (DSN) antennas larger. Increasing their aperture from 64 m to 70 m improved their ability to receive weak signals by 1.4 dB. This would enable them to receive data at a faster data rate while maintaining an acceptably low error rate from a spacecraft that was 4.5 billion kilometers away [16]. The 70-m and 34-m antennas at a given site were then arrayed together to provide even greater sensitivity. Just arraying a 70-m antenna with one 34-m high-efficiency antenna (HEF) improved overall signal strength by another 0.8 dB [17].

After the spacecraft was reprogrammed and the three large DSN antenna aper-

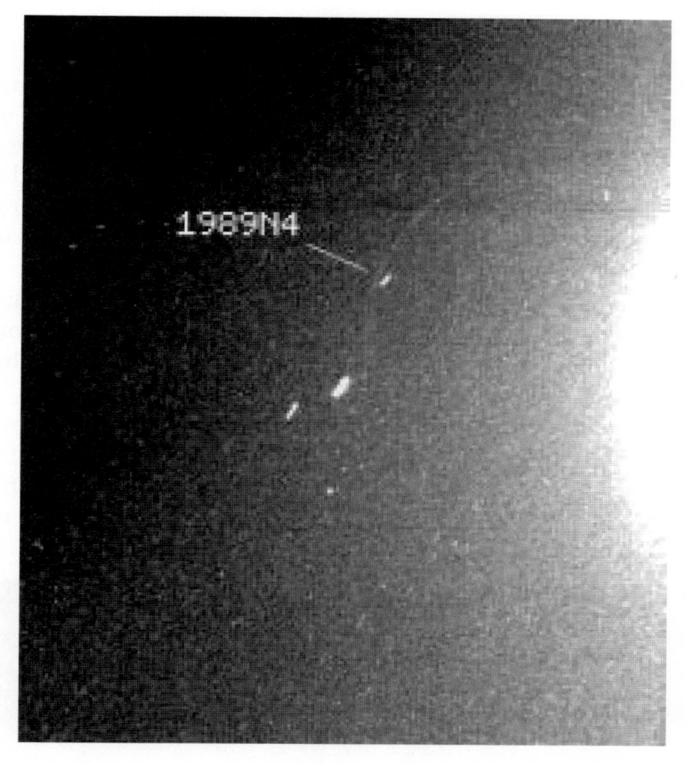


**Figure 5.1.** A 155-s exposure taken with the narrow angle camera captured this image of one of the two ring-arcs and the newly discovered satellite Proteus. (P-34578)

tures increased, negotiations were held between the United States, Japan, and Australia. These negotiations were for use of the Japanese 64-m Usuda Antenna and the Australian 64-m Parkes Radio Telescope. In addition, negotiations were held with the National Radio Astronomy Observatory (NRAO) for the use of their Very Large Array in Socorro, New Mexico. Basically, the Voyager Project wanted to "listen with the entire Pacific Basin" for data being transmitted by Voyager 2 during its historic Neptune encounter.

With the ground antennas enlarged and the spacecraft upgraded, the Voyager Project was ready. The Neptune encounter began on June 5, 1989 [18]. The planet was full of surprises. The planet's upper atmosphere showed unexpected atmospheric features; the planet's magnetosphere was not aligned with its spin axis; Triton, the planet's largest moon, had active geysers; and on August 11, ring astronomers were rewarded with the discovery of two ring-arcs (Figure 5.1). Initial observations indicated that they were 50,000 km and 10,000 km long [19]. Now the questions turned from whether Neptune had rings to trying to understand the mechanism that created the arcs? One immediate possibility was that the Neptune satellites were somehow responsible.

## 5.4 THE RINGS OF NEPTUNE

Data from Voyager 2's encounter revealed that Neptune has five continuous rings. Starting from the innermost ring and working outward, they are now named Galle, Le Verrier, Lassel, Arago, and Adams (Figure 5.2). Each is named after a prominent 19th-Century astronomer who had a major role leading to uncovering the true nature of this trans-Uranian planet. The rings do not all have the same structure. Le Verrier, Arago, and Adams are very narrow and distinct while Galle and Lassell are broad and diffuse.

In addition to the five continuous rings named above, there is a partial ring which shares its orbit with the satellite Galatea. For that reason it is sometimes called the Galatea ring, but it has not been given an official name as yet and very little is known about its detailed characteristics.

Adams (the outermost ring) was also found to have dense clumps of material. These clumps of material are called ring-arcs and are the physical ring features that were responsible for the strange and confusing stellar ring occultation observations. In addition, the strange stellar occultation result obtained by Reitsema *et al.* during the

**Figure 5.2.** These two 591-s exposure images taken by Voyager 2 on August 26, 1989, captured the entire Neptune ring system. Notice that the bright ring-arcs are not in either frame. Unfortunately, the two images were taken 1 hour and 27 min apart during which the ring-arcs were on the other side of the planet. (P-34726)

**Figure 5.3.** This is the first narrow-angle, clear filter image showing Neptune's rings in relatively fine detail. It was taken at 135-degree phase angle when Voyager 2 was 1.1 million km behind Neptune. (P-34712)

May 24, 1981, event was indeed confirmed by Voyager 2 to be a previously unknown moon of Neptune. This satellite is now called Larissa.

In total, Voyager 2 discovered five ring-arcs in the Adams ring. From leading to trailing they are named Courage, Liberté, Egalité 1, Egalité 2, and Fraternité. In Figure 5.3 ring-arcs Fraternité, Egalité, and Liberté can be seen (from left to right).

Neptune's rings were found to be more ill-defined than those of Uranus. Some of this has to do with the much larger number of Earth-based star occultation data sets for the Uranus rings as compared with those for Neptune [20]. Neptune ring data are tabulated in Table 5.3.

In general, planetary rings display very little in the way of azimuthal variations; most of the dramatic variability appears in the radial direction. The Neptune ring-arcs seem to violate that general rule. Many ring scientists believe that the ring arcs are a very short-lived phenomenon and that they will quickly distribute their material around the entire Adams ring. While the Voyager encounter period was very short, especially compared with most geologic periods, the ring-arcs appeared to remain

**Table 5.3.** Neptune ring and ring satellite data from Voyager 2.

| Ring name (satellite) | Radial distance (km) | Width (km) | Optical depth ($\tau$) | Comments |
|---|---|---|---|---|
| Galle | ~41,900 | ~2000 | 0.000 08 | Diffuse, indistinct edges |
| (Naiad) | 48,227 | | | No ring interaction? |
| (Thalassa) | 50,075 | | | No ring interaction? |
| (Despina) | 52,526 | | | May shepherd Le Verrier |
| Le Verrier | 53,200 ± 20 | ~110 | 0.01–0.02 | Narrow, inclined ~0.03° |
| Lassell | 53,200–57,200 | 4,000 | 0.000 15 | Lower dust content |
| Arago | ~57,200 | <100 | ~0.001 | At outer edge of Lassell |
| Unnamed | ~61,950 | <100 | ~0.001 | Diffuse partial ring |
| (Galatea) | 61,953 | | | Probably controls arcs |
| Adams | 62,932 ± 2 | 15–50 | 0.01–0.02 | Arcs have $\tau \sim 0.1$ |
| (Larissa) | 73,548 | | | Largest of ring satellites |

constant over the duration of the encounter period. This is in marked contrast with Saturn's B-ring spokes, one of the only other examples in the solar system of azimuthal structure in ring systems. Saturn's B-ring spokes appeared almost instantaneously, underwent major changes over the course of just one revolution about the planet, and often disappeared within two or three circuits of Saturn. Table 5.4 shows the approximate length and separation between the ring-arcs as determined from Voyager 2 observations.

Hubble Space Telescope observations taken in 2005 have revealed that there have been changes in Neptune's ring system since the Voyager 2 encounter. The Adams and Le Verrier rings are still at their expected locations but the ring-arc Liberté has almost completely vanished and Courage appears to have moved an additional 8° ahead of its prior position [21]. This is in contrast with ring-arcs Egalité and Fraternité which do not appear to have changed at all. Clearly, something is going on that cannot be explained by Voyager data alone.

**Table 5.4.** Neptune ring-arc dimensions within the Adams ring.

| Arc name | Approximate arc length (deg) | Separation from leading arc (deg) | Comments |
|---|---|---|---|
| Courage | 1.0 ± 0.1 | 0.0 | Least optically thick of arcs |
| Liberté | 4.1 ± 0.1 | 7.3 | Well-defined arc length |
| Egalité | 1.0 ± 0.5 | 19.6 | The boundary between the two components of the Egalité arc is poorly defined |
| Egalité 2 | 3.0 ± 0.5 | 22.4 | |
| Fraternité | 9.6 ± 0.1 | 33.2 | Longest of the arcs |

## 5.5 NOTES AND REFERENCES

[1] *http://solarsystem.nasa.gov/planets/profile*—Uranus.

[2] *http://solarsystem.nasa.gov/planets/profile*—Neptune.

[3] Elliot, J., Kerr, R., 1984, *Rings—Discoveries from Galileo to Voyager*, The Massachusetts Institute of Technology Press, p. 182.

[4] *Ibid.*

[5] Guinan, E. F., Harris, C. C., Maloney, F., 1982, "Evidence for a ring system of Neptune", *Bulletin of the American Astronomical Society* **14**, 658 (abstract only).

[6] Reitsema, H. J., Hubbard, W. B., Lebofsky, L. A., Tholen, D. J., 1982, "Occultation by a possible third satellite of Neptune", *Science* **215**, 289–291.

[7] Hubbard, W. B., Frecker, J. E., Gehrels, J. A., Gehrels, T., Hunten, D. M., Lebofsky, L. A., Smith, B. A., Tholen, D. J., Vilas, F., Zellner, B., Avey, H. P., Mottram, K., Murphy, T., Varnes, B., Carter, B., Nielsen, A., Page, A. A., Fu, H. H., Wu, H. H., Kennedy, H. D., Waterworth, M. D., Reitsema, H. J., 1985, "Results from observations of the 15 June 1983 occultation by the Neptune system", *The Astronomical Journal* **90**, 655–667.

[8] Cruikshank, D. P. (ed), 1995, *Neptune and Triton*, The University of Arizona Press, pp. 734–735.

[9] Lissauer, J. J., 1985, "Shepherding model for Neptune's arc ring", *Nature* **318**, 544–545.

[10] Sicardy, B., Roques, F., Brahic, A., Bouchet, P., Maillard, J. P., Perrier, C., 1986, "More dark matter around Uranus and Neptune?", *Nature* **320**, 729–731.

[11] Goldreich, P., Tremaine, S., Borderies, N., 1986, "Towards a theory for Neptune's arc rings", *The Astronomical Journal* **92**, 490ff.

[12] Elliot, J., Kerr, R., 1984, *Rings—Discoveries from Galileo to Voyager*, The Massachusetts Institute of Technology Press, p. 186.

[13] Miner, E. D., Ingersoll, A., Kurth, W., Esposito, L., Johnson, T., 1987, *Science Objectives and Preliminary Sequence Designs for the Voyager Neptune Encounter: Report of the Neptune Science Working Group*, JPL Publication D-4607/PD 1618-66. National Aeronautics and Space Administration, Jet Propulsion Laboratory, California Institute of Technology, Pasadena.

[14] Miner, E. D., Wessen, R. R., 2002, *Neptune: The Planet, Rings and Satellites*, Springer-Praxis, p. 147.

[15] *Ibid.*, p. 153.

[16] Ludwig, R., Taylor, J., 2002, *Voyager Telecommunications*, Design and Performance Summary Series, Jet Propulsion Laboratory, p. 34.

[17] *Ibid.*, p. 34.

[18] Miner, E. D., Wessen, R. R., 2002, *Neptune: The Planet, Rings and Satellites*, Springer-Praxis, p. 163.

[19] *Ibid.*, p. 174.

[20] Miner, E. D., Wessen, R. R., 2002, *Neptune: The Planet, Rings and Satellites*, Springer-Praxis, p. 240.

[21] De Pater, I., Gibbard, S. G., Chiang, E., Hammel, H. B., Macintosh, B., Marchis, F., Martin, S. C., Roe, H. G., Showalter, M., 2005, "The dynamic Neptunian ring arcs: Evidence for a gradual disappearance of Liberté and resonant jump of Courage", *Icarus* **174**, 263–272.

## 5.6 BIBLIOGRAPHY

Elliot, J., Kerr, R., 1984, *Rings—Discoveries from Galileo to Voyager*, The Massachusetts Institute of Technology Press.

Miner, E. D., Wessen, R. R., 2002, *Neptune: The Planet, Rings and Satellites*, Springer-Praxis.

Miner, E. D., Ingersoll, A., Kurth, W., Esposito, L., Johnson, T., 1987, *Science Objectives and Preliminary Sequence Designs for the Voyager Neptune Encounter: Report of the Neptune Science Working Group*, JPL Publication D-4607/PD 1618-66. National Aeronautics and Space Administration, Jet Propulsion Laboratory, California Institute of Technology, Pasadena.

Porco, C. C., Nicholson, P. D., Cuzzi, J. N., Lissauer, J. J., Esposito, L. W., 1995, "Neptune's ring system", in *Neptune and Triton*, edited by D. P. Cruikshank, The University of Arizona Press, pp. 703–804.

## 5.7 PICTURES AND DIAGRAMS

Figure 5.1 *http://photojournal.jpl.nasa.gov/jpeg/PIA02200.jpg*
Figure 5.2 *http://photojournal.jpl.nasa.gov/jpeg/PIA01997.jpg*
Figure 5.3 *http://photojournal.jpl.nasa.gov/jpeg/PIA02207.jpg*

# 6

# Present knowledge of the Jupiter ring system

## 6.1 INTRODUCTION

In Chapter 4, we covered the discovery of the Jupiter ring system, the sole example of a ring system first discovered by a spacecraft. Scientists preparing the Jupiter encounter observations of Voyager 1 did not devote much time or effort to ring observations, primarily because no ring of Jupiter had ever been detected. The single observation that was added to the Voyager 1 encounter sequence was a long-exposure image designed to search for a possible tenuous ring near Jupiter's equatorial plane.

The Pioneer 11 spacecraft would not arrive at Saturn to reveal new rings there until September 1, 1979, shortly after the completion of Voyager 2's Jupiter encounter sequence (Chapter 2). Initial attempts to detect Neptune's rings by means of stellar occultation measurements did not begin until a year later, although possible evidence for detection of a partial ring was thereafter found in a 1968 measurement (Chapter 5). It was not until the Voyager 1 detection of a Jupiter ring system that ring scientists came to realize that planetary ring systems are a seemingly ubiquitous characteristic of giant planets like Jupiter, Saturn, Uranus, and Neptune. Now it appears likely that rings will eventually be sighted around giant planets discovered in other solar systems within the Orion Spur of our Milky Way Galaxy.

Fortunately, the Voyager 1 discovery came at a time when it was still possible to make slight alterations to the Voyager 2 Jupiter encounter, thereby enabling capture of the Voyager mission's best images of Jupiter's ring system. Armed with these defining data, the Galileo mission, a long-term Jupiter orbital mission, was able to plan for more detailed imaging of Jupiter's ring system, and it is primarily the results and analysis of these data that we discuss in this chapter.

Pioneer 11 also played a pivotal role in the discovery of the Jupiter ring system. Its 1974 data revealed an unexpectedly low energetic charged particle population planetward of Jupiter's little satellite, Amalthea, which orbits Jupiter at a range of 181,400 km. Jupiter's equatorial radius is 71,492 km, so Amalthea orbits at a scant

1.54 Jupiter radii above the cloud tops. Noting the paucity of energetic particles, Pioneer magnetometer scientists Acuña and Ness [1] suggested that an unseen ring or undiscovered satellite might be responsible for absorbing some of the charged particle flux inward of Amalthea's orbit. In hindsight, their suggestion proved to be prophetic. However, other explanations for the reduced flux of energetic particles were also suggested, so there was no unanimity of opinion on the cause or causes, and little attention was given to the suggestion until after the Voyager 1 Jupiter encounter.

It is interesting to note that, had Voyager 1 not made its serendipitous discovery, Voyager 2 might not have devoted any of its precious time to capturing additional ring data at Jupiter, and Galileo, with its crippled antenna and drastically reduced data return capability, might not have deemed a Jupiter ring search of high potential value. But for the unexpected discovery, we might even now know little or nothing about the ring system of the solar system's largest planet.

## 6.2 RING RADIAL CHARACTERISTICS

There is very little visible radial structure within the Jupiter ring system other than the boundaries between the different ring portions. The Main ring is both the narrowest and the thinnest of the rings, with a radial extent of 7,200 km and a vertical extent of less than 30 km [2]. Each of the other rings is brightest near its outer edge, which also likely corresponds to the location of the respective primary source of the ring's material. The source of material in the Halo ring is the Main ring; sources for the inner and outer Gossamer rings are Amalthea and Thebe, respectively. The general dimensions of the Jupiter rings are given in Table 6.1. An image and diagram of the rings was given in Figures 4.4 and 4.5. Detailed information on the individual rings is given in the following sections of this chapter.

## 6.3 MAIN RING

All of Jupiter's rings appear to be composed primarily of dust-sized particles, as clearly demonstrated by the fact that they are much brighter in Voyager 2 and Galileo images looking back in the general direction of the Sun than they are in images with the Sun and the ring in opposite directions from the spacecraft. This characteristic is typical of particles in the size range of about 1 micrometer in radius, much like dust on an automobile windshield appears bright and obscures a driver's vision when driving toward the Sun but is scarcely visible when traveling away from the direction of the Sun.

Both their relative positioning and brightness gradation within the Main ring make it apparent that the small satellites Metis and Adrastea (see Figure 6.1) provide much or all of the material in the ring. Jupiter's immense gravity focuses the paths of many meteoroids that approach the planet, and while most of these meteoroids fall into the atmosphere and are burned up, satellites close to Jupiter undergo heavy bombardment. These meteoroids penetrate deep into the respective satellites, becom-

**Table 6.1.** Jupiter ring dimensions ($1R_J = 71{,}492$ km, $1M_J = 18.987 \times 10^{26}$ kg).

| Ring region | Boundary (km) | Boundary ($R_J$) | Velocity (km/s) | Period (hr) | Mass ($M_J$) |
|---|---|---|---|---|---|
| Halo ring | 100,000 | 1.40 | 36* | 4.9* | ** |
| | 122,000 | 1.71 | 32* | 6.6* | |
| Main ring | 122,000 | 1.71 | 32 | 6.6 | $\sim 0.0001 \times 10^{-10}$ |
| *(Metis* | *128,000* | *1.79* | *31.50* | *7.09* | $0.6 \times 10^{-10}$*)* |
| *(Adrastea* | *129,000* | *1.80* | *31.38* | *7.16* | $0.04 \times 10^{-10}$*)* |
| | 129,200 | 1.81 | 31.32 | 7.17 | |
| Inner Gossamer ring | 129,200 | 1.81 | 31.32 | 7.17 | ** |
| *(Amalthea* | *181,400* | *2.54* | *26.57* | *11.96* | $11 \times 10^{-10}$*)* |
| | 182,000 | 2.55 | 26.57 | 12.02 | |
| Outer Gossamer ring | 182,000 | 2.55 | 26.57 | 12.02 | ** |
| ... *(Thebe* | *221,900* | *3.10* | *23.92* | *16.19* | $8 \times 10^{-10}$*)* |
| | 224,900 | 3.15 | 23.89 | 16.43 | |

* The velocities and periods given for the Halo ring are for equatorial circular orbits of neutral particles. In the Halo ring, motions of the particles are very likely affected by the magnetic field of Jupiter, which rotates once every 9.925 hr, independent of distance from the planet.

** Most of the mass of Jupiter's ring system is probably in the Main ring. No estimates are available for the mass of the other three rings, but they are much smaller than that of the Main ring.

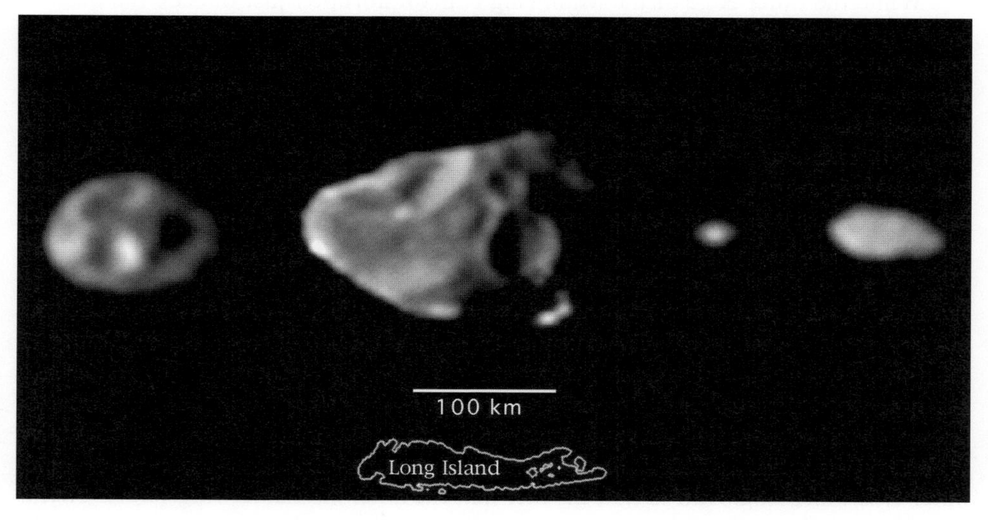


**Figure 6.1.** The four small satellites that are the primary sources for the Jupiter ring system are shown to scale and with their north poles approximately at the top. From left to right they are Thebe, Amalthea, Adrastea, and Metis. Shown to the same scale is Long Island, New York, which is approximately 190 km in length. (PIA01625)

ing superheated in the process and then exploding, creating craters and debris. If the satellite is large, most of the debris falls back to the satellite surface, but the low gravity of smaller moonlets permits much of the debris to escape and go into orbit around Jupiter. Adrastea, with a mean diameter of only 20 km probably provides much more of the Main ring material than does 43-km-wide Metis; Metis's surface gravity is three times that of Adrastea. Even smaller unseen moons may also provide some of the debris that forms the Main ring.

Once separated from their source moonlets, material in the Main ring continues to undergo bombardment, breaking it into smaller and smaller pieces. In denser rings, mutual collisions between the ring particles cause them to spread radially both toward and away from the planet. But, for the tenuous rings of Jupiter, including the Main ring, interparticle collisions are relatively rare, and other effects dominate. In the Main and Gossamer rings, the primary force appears to be Poynting–Robertson drag. This is an effect caused by sunlight, which comes from a single direction and heats the ring particle. If the particle is small (i.e., dust-sized or smaller), the entire particle is warmed and re-emits the energy in all directions. Although the energies involved are small, they are sufficient to cause small particles to slow down, thereby spiraling planetward over time. Most of the Main ring particles appear to be lost to the Halo ring (see Section 6.5), where other forces due to interaction with the Jupiter magnetic field and electrically charged particles within that field add to the Poynting–Robertson drag. The combined effect is to cause the dust-sized and smaller ring particles to fall into Jupiter's atmosphere within 1,000 years or less [3]. The Main ring (and the other Jupiter rings) must therefore be continually replenished or they would disappear in extremely short times compared with the 4.5-billion-year age of Jupiter.

While the outer edge of the Main ring cuts off abruptly just outside the orbit of Adrastea, its inner edge is much less distinct. At its outer edge the thickness of the Main ring is less than 30 km (as mentioned in Section 6.2), comparable with the diameters of its larger source moonlets. The ring is also brightest (and densest?) at its outer edge. However, as material moves inward, the ring becomes thicker, possibly an indication that electromagnetic forces play more of a role.

## 6.4 GOSSAMER RINGS

The inner Gossamer ring has as its source the satellite Amalthea (see Figure 6.1). Amalthea has a mean diameter of 167 km, but—more importantly—it has an orbital inclination to Jupiter's equator of 0.388°. With its mean orbital radius of 181,400 km, an inclination of 0.388° leads to excursions above and below Jupiter's equator of 1,230 km, or a full thickness of 2,460 km. If one then adds Amalthea's mean diameter of 172 km, the result is remarkably close to the 2,600-km estimated thickness of the inner Gossamer ring. Add to that the fact that the inner Gossamer ring's outer edge is essentially at the orbit of Amalthea, especially considering that the eccentricity of the orbit carries Amalthea more than 800 km farther than its mean distance, and the correspondence is unmistakable.

A similar situation exists for the outer Gossamer ring. Its estimated thickness is

8,800 km. Thebe's mean orbital radius is 221,900 km and its orbital inclination is 1.070°, leading to excursions above and below Jupiter's equator of 4,140 km, or a full thickness of 8,280 km. Thebe's mean diameter is about 100 km (Figure 6.1). While the match (8,380 km versus 8,800 km) is not as close as for the inner Gossamer ring, the bright part of the outer Gossamer ring again extends to the orbit of Thebe. A very faint outward extension of the outer Gossamer ring also has the same 8,800-km thickness. A faint outward extension of the inner Gossamer ring would probably be hidden in the glare of the outer Gossamer ring.

Although Table 6.1 indicates that the inner edges of the outer and inner Gossamer rings are coincident with the outer edges of the inner Gossamer ring and the Main ring, respectively, each of the Gossamer rings actually extends much further inward, albeit at reduced brightness levels. It is likely that the outer Gossamer ring surrounds the inner Gossamer ring, which in turn surrounds the Main ring. The orbital motions of Amalthea and Thebe are such that these satellites spend more time near their extreme upper and lower limits than at any other intermediate distance above or below Jupiter's equatorial plane. The Gossamer rings are consequently brightest at their upper and lower extremes rather than being disks of uniform brightness. The orbits of Amalthea and Thebe also precess (change the orientation of their orbits while maintaining the same tilt) rapidly, wobbling like slowly spinning hoops which return their orientation to the original position in only a few months. Therefore, over periods as short as a few months, the Gossamer rings form relatively uniform disks of dusty material, densest at their upper and lower surfaces and least dense at their centers.

Because of their association with the two satellites of Jupiter, the outer Gossamer ring is sometimes called the Thebe Gossamer ring, and the inner Gossamer ring is sometimes called the Amalthea Gossamer ring.

## 6.5 HALO RING

Four small satellites (possibly aided by smaller unseen moonlets) provide the dusty material that makes up the Main and Gossamer rings of Jupiter. Material in those rings is acted upon by Poynting–Robertson drag (described in Section 6.3) and slowly moves planetward. However, when it reaches a radial distance of 122,150 km from the center of Jupiter, it completes precisely three orbits of Jupiter for every two rotations of the planet's magnetic field. (Note that the planetary magnetic field has a rotation period of 9.924 hr.) Hence, any tiny ring particles that have acquired an electrical charge will experience slight vertical forces that tend to push the particle out of the ring plane. At the radius of the so-called 3:2 *Lorentz resonance*, those forces will be in the same direction every third orbit of the ring particles, and the near-equatorial orbits of particles moving inward from the Main ring are very quickly spread vertically to a thickness approaching 20,000 km.

Poynting–Robertson drag then continues to move the tiny particles further inward until they reach the 2:1 Lorentz resonance, which occurs when the natural orbit period of the particles is half that of the Jupiter rotation period. That occurs near a radial distance of 100,450 km, where the orbital period is 4.962 hr and is strong enough for sub-micrometer particles such that it rivals the gravity of Jupiter. At that point, the vertical forces on charged particles are much stronger than for the 3 : 2 Lorentz resonance, and Halo ring particles are carried completely out of the Jupiter ring, the vast majority quickly entering the atmosphere of Jupiter and becoming a part of the planet itself. From inspection of Table 6.1, it is clear that the boundaries of the Halo ring correspond closely with these two resonance radii. The electromagnetic process briefly discussed here is described in more detail in several good scientific articles [4]. There are other Lorentz resonances within the ring system. The 1 : 1 synchronous orbital radius occurs midway through the inner Gossamer ring, for example. However, there seems little in the way of visible effects from this or the more distant Lorentz resonances.

There are several possible reasons for that fact: first, the particles might be too large to be easily deflected (in other words, the charge-to-mass ratio is too small); second, the particles may not yet have acquired an electrical charge (Lorentz forces work only on electrically charged particles); or, third, the Gossamer rings are already very faint, and vertical deflection of a small fraction of their particle populations may be entirely undetectable.

How do the particles become electrically charged? Jupiter's magnetic field is populated with charged particles, which circle the planet in the same amount of time as the planet's rotation period (9.924 hr). The magnetic axis of Jupiter is tilted by 10° to the rotation axis. Therefore, in general, orbiting ring particles will cross the magnetic equator twice each rotation of the planet. During the ensuing collisions, the ring particles acquire some of the electrical charge of the magnetospheric particles. At the 1 : 1 resonance distance in the inner Gossamer ring, however, the orbiting ring particles maintain their position with respect to the planet's magnetic equator, and few collisions between magnetospheric particles and ring particles occur. This charging mechanism therefore becomes more efficient the farther the ring particles are from the 1 : 1 resonance radius. It is perhaps noteworthy that the Jupiter ring system extends from the 2 : 1 Lorentz resonance (the inner boundary of the Halo ring) to the 1 : 2 Lorentz resonance at the outer edge of the outer Gossamer ring (see Figure 6.2).

For those ring particles closer in to the planet, a second charging mechanism becomes important. Jupiter's ionosphere electron population peaks near altitudes that vary from a few hundred to a few thousand kilometers above the cloud tops, both well planetward of the known rings [5]. However, the ionosphere falls off very slowly with altitude, and there are still remnants of this electron population near the inner edge of the Jupiter rings. These also move with the magnetic field and can serve as an electrical charge source for close-in ring particles.

The third mechanism for charging ring particles operates relatively equally in all parts of the rings. Like the Poynting–Robertson effect, it depends on sunlight. It is known as the *photoelectric effect*. When sunlight impinges on atoms and molecules near the surfaces of ring particles, occasionally enough energy is added to those atoms or molecules to free some of the outer electrons, which escape into space, leaving the ring particle with a small positive charge. For large particles, this has a negligible effect on their motions, but for tiny particles, this can serve over time to build up a relatively

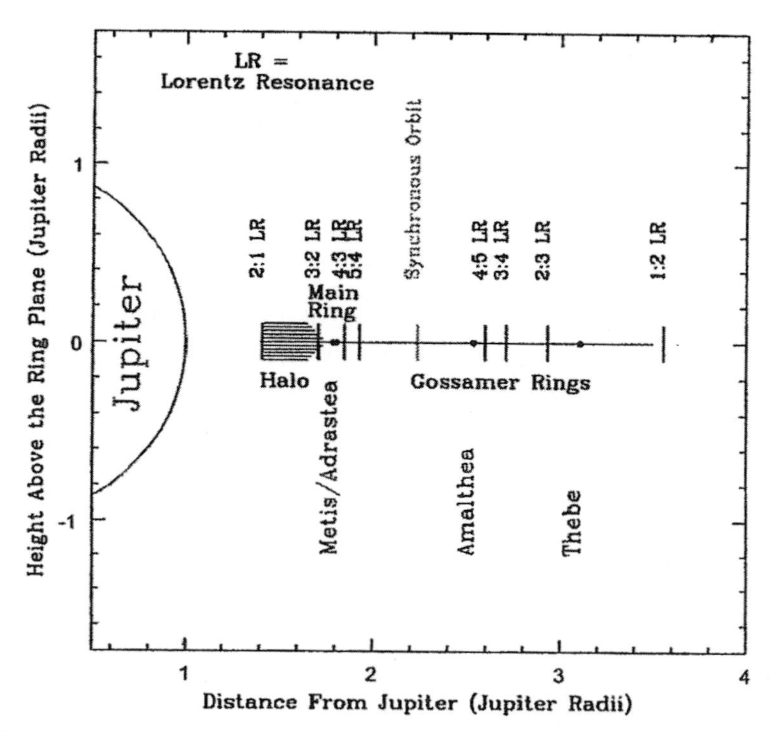


**Figure 6.2.** Lorentz resonances, which mark ring radii where enhanced interactions with the Jupiter magnetosphere occur, are shown in this diagram from Burns *et al.* (2004) [4]. The 2:1 and 3:2 Lorentz resonances bound the Halo ring, and the 1:2 resonance may bound the outer edge of the Gossamer rings.

large charge-to-mass ratio for the particles, whose motions are then more readily perturbed by interaction with Jupiter's magnetic field.

## 6.6 RING-PARTICLE PROPERTIES

Voyager observations of Jupiter's Galilean satellites (Io, Europa, Ganymede, and Callisto) reveal that their densities decrease with increasing distance from Jupiter. Their respective densities, in grams per cubic centimeter, are 3.57, 3.04, 1.94, and 1.86 [6]. The mean density of Earth's Moon, composed mainly of rocky materials, is 3.34 g/cm$^3$; the density of pure water is 1.00 g/cm$^3$. This range of densities of the four Galilean satellites has been interpreted as an indication that heat from Jupiter during the planet's formative stages was responsible for driving volatile matter (like water) away from the inner portions of the Jupiter satellite system. No evidence is found for water at Io. Of course, that may be at least partly due to Io's volcanic activity. Europa was heated enough to allow the less dense materials (like water) to rise to the surface, so in spite of its relatively high density, it has a surface completely covered with water ice and perhaps a deep water ocean beneath the icy crust. Ganymede and Callisto are probably more like Earth's Moon, although they probably have more water ice (and possibly liquid water) in their interiors.

If indeed Jupiter's heat did drive away volatiles in the inner satellite system, the four source satellites for Jupiter's rings (Metis, Adrastea, Amalthea, and Thebe) would likely also be waterless objects. Although their sizes are fairly well determined, only Amalthea's mass has been measured (during the November 2002 close approach of the Galileo spacecraft). From radio data obtained during that encounter, Anderson *et al.* [7] have determined an Amalthea density of $0.857 \pm 0.099\ \mathrm{g/cm^3}$, which is certainly less than that of water. Such a low density suggests a high amount of porosity. Amalthea (and the other Jupiter ring moons?) may be "rubble piles" rather than well-consolidated natural satellites. If Amalthea were composed entirely of rocky materials, loose packing alone would likely be insufficient to yield so low a density; the low density led Anderson *et al.* to conclude that a substantial fraction of Amalthea's composition is water ice. That conclusion in turn requires abandonment of the theory that Jupiter's formative heat drove out volatiles or, alternatively, an origin for Amalthea in colder regions of the solar system and subsequent capture by Jupiter after the planet had cooled. The general compositions and origins of the other three ring moons are likely similar to those of Amalthea. Whatever their internal composition, all three moons are dark objects, reflecting only 5–9% of the sunlight incident on their surfaces. The low reflectivity testifies to the effects of the heavy meteoroid bombardment they have undergone; most meteoroids are dark.

Direct measurements of the composition of Jupiter's ring particles are difficult to obtain, either from Earth or from spacecraft. The difficulty is due to their closeness to the planet and to their inherent faintness. Multicolor imaging of the rings from Voyager and Galileo, especially the main ring, confirms that they reflect red light more efficiently than blue light. A reddish hue is more characteristic of silicate (rocky) material than of icy material. In the early 1980s, direct measurements of their reflectivity were attempted from Earth by Smith and Reitsema [8] and Neugebauer *et al.* [9]. The Galileo near-infrared mapping spectrometer obtained a few spectral observations of Jupiter's rings, but all from very high ($\sim 180^\circ$) phase angles. Perhaps the best data set on Jupiter's rings to date is from Cassini observations during a 6-month period surrounding Cassini's closest approach on December 30, 2000.

An attempt to integrate Earth-based and spacecraft (Galileo and Cassini) observations was made by Throop *et al.* [10], and the discussion from this point forward in this chapter is taken mostly from that treatise. It seems that the phase curve of Jupiter's ring system is dominated at low phase angles by the larger (non-dust) ring particles, whereas the dust component is dominant at high phase angles (see Figure 6.3). The model fit seems to show that, for unknown reasons, the particle population peaks at particle sizes near 15 micrometers in radius and that the particles are likely irregular in shape rather than spherical. The data confirm that the larger ring particles (from which the dust-sized grains are derived) are very red (much like Amalthea), at least over the wavelength range from 0.4 micrometers in the blue to 2.5 micrometers in the near infrared. No evidence for water or other ices is found in the data, although

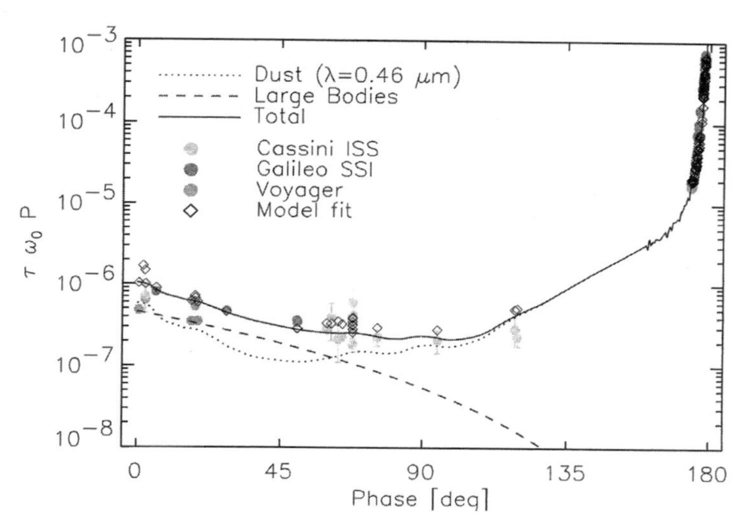


**Figure 6.3.** The relative brightness of the Jupiter rings at phase angles from 0° to 180°, taken from Throop *et al.* (2004) [10]. The jovian ring particle population has two components: a dust component which dominates the phase curve at high and intermediate phase angles, and a larger particle population which is brighter in backscatter (i.e., low phase angles, including all Earth-based observations). The diamonds are a best-fit mathematical model of ring particle light-scattering behavior.

there are perhaps weak absorptions near 0.8 micrometers (red light) and 2.2 micrometers (near infrared radiation); no suggestions for the source of these possible absorptions was put forward.

## 6.7 NOTES AND REFERENCES

[1] Acuña, M. H., Ness, N. F., 1976, "The complex main magnetic field of Jupiter", *Journal of Geophysical Research* **81**, 2917–2922.

[2] Burns, J. A., 1999, "Planetary rings", Chapter 16 in *The New Solar System* (4th Edition), edited by Beatty, Petersen, and Chaikin, pp. 221–240. The Main ring thickness is given in table 1 on p. 227.

[3] *Ibid.*, pp. 228–229.

[4] Burns, J. A., Schaffer, L. E., Greenberg, R. J., Showalter, M. A., 1985, "Lorentz resonances and the structure of the Jovian ring", *Nature* **316**, 115–119; Schaffer, L., Burns, J. A., 1992, "Lorentz resonances and the vertical structure of dusty rings: Analytical and numerical results", *Icarus* **96**, 65–84; Burns, J. A., Simonelli, D. P., Showalter, M. R., Hamilton, D. P., Esposito, L. W., Porco, C. C., Throop, H., 2004, "Jupiter's Ring–Moon System", in *Jupiter. The planet, satellites, and magnetosphere*, edited by Bagenal, Dowling, and McKinnon, Cambridge Planetary Science Series, pp. 241–262; De Pater, I., Showalter, M. R., Burns, J. A., Nicholson, P. D., Liu, M., Hamilton, D. P., Graham, J. R., 1999,

"Keck infrared observations of Jupiter's ring system near Earth's 1997 ring plane crossing", *Icarus* **138**, 214–223.

[5] See, for example, Yelle, R. V., Miller, S., 2004, "Jupiter's thermosphere and ionosphere", in *Jupiter. The planet, satellites and magnetosphere*, edited by Bagenal, Dowling, and McKinnon, Cambridge Planetary Science Series, pp. 185–218.

[6] Burns, J. A., 1986, "Some background about satellites", in *Satellites*, edited by Burns and Matthews, pp. 1–36.

[7] Anderson, J. D., Johnson, T. V., Schubert, G., Asmar, S., Jacobson, R. A., Johnston, D., Lau, E. L., Lewis, G., Moore, W. B., Taylor, A., Thomas, P. C., Weinwurm, G., 2005, "Amalthea's density is less than that of water", *Science* **308**, 1291–1293.

[8] Smith, B. A., Reitsema, H. J., 1980, "CCD observations of Jupiter's ring and Amalthea", in *Satellites of Jupiter*, volume 57 of the International Astronomical Union's Colloquium Series.

[9] Neugebauer, G., Becklin, E. E., Jewitt, D., Terrile, R., Danielson, G. E., 1981, "Spectra of the Jovian ring and Amalthea", *Astronomical Journal* **86**, 607–610.

[10] Throop, H. B., Porco, C. C., West, R. A., Burns, J. A., Showalter, M. R., Nicholson, P. D., 2004, "The jovian rings: New results derived from Cassini, Galileo, Voyager, and Earth-based observations", *Icarus* **172**, 59–77.

## 6.8 BIBLIOGRAPHY

*Jupiter. The planet, satellites and magnetosphere*, 2004, Cambridge University Press, edited by Fran Bagenal, Timothy Dowling and William McKinnon. Note especially the following chapters: Krüger, H., Horányi, M., Krivov, A. V., Graps, A. L., "10. Jovian dust: Streams, clouds and rings", pp. 219–240; Burns, J. A., Simonelli, D. P., Showalter, M. R., Hamilton, D. P., Esposito, L. W., Porco, C. C., Throop, H., 2004, "11. Jupiter's Ring–Moon System", pp. 241–262.

*Planetary Rings*, 1984, University of Arizona Press, edited by Richard Greenberg and André Brahic. Note especially the following chapters: Burns, J. A., Showalter, M. R., Morfill, G. E., "The ethereal rings of Jupiter and Saturn", pp. 200–272; Grün, E., Morfill, G. E., Mendis, D. A., "Dust–magnetosphere interactions", pp. 275–332; Mignard, F., "Effects of radiation forces on dust particles", pp. 333–366.

## 6.9 PICTURES AND DIAGRAMS

Figure 6.1 *http://photojournal.jpl.nasa.gov/jpeg/PIA01625.jpg*
Figure 6.2 Fig. 11.14 from Burns *et al.* (2004) [4].
Figure 6.3 Fig. 4 from Throop *et al.* (2004) [10].

# 7

# Present knowledge of the Uranus ring system

## 7.1 INTRODUCTION

Voyager 2 provided the first resolved images of the rings of Uranus and of many small satellites which might gravitationally influence them. It also obtained data from a variety of sensors and over a variety of viewing and illumination conditions which help to determine the reflective properties and composition of the ring particles as well as their particle size distribution. However, continuing Earth-based measurements of stellar ring occultations (i.e., blockage of starlight by the rings as viewed from Earth) have provided the longest time base and most accurate data on ring radii, shapes, inclinations, widths, *optical depths* (transparency), *precession rates* (how fast the orbit changes orientation), and dynamic stability, all data that were not obtained by Voyager 2 in the short time period during which it was in the vicinity of Uranus. This chapter will follow up the discussions of Chapter 3 on the discovery of the Uranus rings with a discussion of what we really know about the rings 30 years after their initial discovery.

Except for three broad and indistinct rings (the innermost and the two outermost), none of which have yet been given official names by the Nomenclature Committee of the International Astronomical Union, the rings of Uranus are all narrow, relatively sharp-edged bands of intrinsically dark particles. Early Earth-based attempts to image the rings between the 1977 discovery and the 1986 Voyager 2 encounter were hampered by their narrowness, their low reflectivity, their great distance from Earth, and their close proximity to Uranus.

Even following the Voyager 2 encounter and more than 20 years of additional theoretical and observational data since 1986, ring scientists are hard-pressed to explain the mechanisms which confine the rings. Cordelia is about 1,400 km interior to the Epsilon ring and Ophelia is about 2,600 km exterior to the Epsilon ring; it is tempting to assume that these two small satellites gravitationally impede radial spreading in the Epsilon ring. The case for shepherding of the Epsilon ring by Cordelia (inner edge) and Ophelia (outer edge) is in fact stronger than for any other Uranus ring. Three other rings may possibly be gravitationally constrained by a known satellite at either a ring inner edge or an outer edge, but not both. For the vast majority of the Uranus rings, no known satellites can provide the gravitational forces needed to confine the ring particles. A thorough search by Voyager 2 and subsequent searches by Earth-based observers have not been successful in locating satellites closer to the planet than Cordelia. Either the satellites which might shape the rings of Uranus are too small to have been discovered as yet or else some mechanism other than gravitational shepherding is acting within this unique ring system.

## 7.2 RING DIMENSIONS, SHAPES, AND INCLINATIONS

Table 7.1 provides an overview of the mean radial distances, velocities, periods, and masses of the 13 known rings and the 14 inner satellites of Uranus [1]. Their widths, orbital eccentricities and inclinations, and *equivalent depths* are given in Table 7.2 [2]. Because the eccentric rings have widths that vary with orbital longitude, the concept of equivalent depth (physical width multiplied by optical depth) is important because it remains essentially constant around the ring and is a measure of the number of particles passing a given point in the rings per unit time.

[The following discussion is restricted to the nine brightest rings; i.e., it includes neither the Lambda ring nor the three broad rings. None of these dusty rings is seen in stellar occultations from Earth.]. From Table 7.2, it is seen that all of the rings except Eta, Gamma, and Delta depart measurably from circularity. Widths of the Epsilon, Delta, Beta, and Alpha rings vary such that they are widest at their most distant excursions from the planet and narrowest when they are closest to the planet. The Eta, Gamma, Delta, and Epsilon rings are confined (within measurement uncertainties) to Uranus's equatorial plane; the five inner rings have small but measurable inclinations. The Epsilon ring and the five inner rings (Six, Five, Four, Alpha, and Beta) have orbits which precess (change the orientation in space of their elliptical orbits) uniformly with time. One of the more interesting results of the Earth-based studies of the rings is their tendency to behave almost like rigid bodies. For example, precession rates of the inner and outer edges of the Epsilon ring would be expected on theoretical grounds to be slightly different, but they are identical.

The longitude of the periapsis (closest point to the planet within its elliptical orbit) of the Epsilon ring precesses at a rate of 1.363 25° per day. In other words, the elliptical orbit shape of the Epsilon ring "revolves" around Uranus every 264 days. This precession is due in part to the non-spherical shape of the planet. The other rings with non-circular shapes (eccentricities greater than 0, see Table 7.2) also precess around the planet at rates which increase with decreasing distance from Uranus. For example, the orbit of the most eccentric of the Uranus rings, the Five ring, precesses at a rate of 2.671 51° per day, completing a full 360° precession in only 121 days.

In a similar fashion, the point at which inclined ring orbits cross the Uranus equator also moves around the planet. The rates of this so-called *nodal regression* are very close in magnitude to those of the *apsidal precession* rates for each ring discussed

**Table 7.1.** Uranus ring dimensions ($1R_U = 25{,}559$ km, $1M_U = 8.6849 \times 10^{25}$ kg).

| Ring region | Mean distance (km)* | Mean distance ($R_U$)* | Velocity (km/s) | Period (hr) | Mass ($M_U$) |
|---|---|---|---|---|---|
| $\zeta$ (Zeta) ring | ~39,600 | ~1.55 | ~12.11 | 5.71 | ?? |
| 6 (Six) ring | 41,837.2 | 1.63689 | 11.780 | 6.1988 | ?? |
| 5 (Five) ring | 42,234.8 | 1.65244 | 11.724 | 6.2875 | ?? |
| 4 (Four) ring | 42,570.9 | 1.66559 | 11.677 | 6.3628 | ?? |
| $\alpha$ (Alpha) ring | 44,718.4 | 1.74962 | 11.393 | 6.8508 | $\sim 3 \times 10^{-10}$ |
| $\beta$ (Beta) ring | 45,661.0 | 1.78650 | 11.274 | 7.0688 | $\sim 2 \times 10^{-10}$ |
| $\eta$ (Eta) ring | 47,175.9 | 1.84577 | 11.091 | 7.4239 | ?? |
| $\gamma$ (Gamma) ring | 47,626.9 | 1.86341 | 11.037 | 7.5315 | ?? |
| $\delta$ (Delta) ring | 48,300.1 | 1.88975 | 10.961 | 7.6911 | ?? |
| *(Cordelia* | *49,752* | *1.9466* | *10.799* | *8.0408* | *$5.2 \times 10^{-10}$)* |
| $\lambda$ (Lambda) ring | 50,023.9 | 1.95719 | 10.770 | 8.1069 | ?? |
| $\varepsilon$ (Epsilon) ring | 51,149.3 | 2.00123 | 10.650 | 8.3823 | $\sim 560 \times 10^{-10}$ |
| *(Ophelia* | *53,764* | *2.1035* | *10.387* | *9.0338* | *$6.2 \times 10^{-10}$)* |
| *(Bianca* | *59,165* | *2.3148* | *9.901* | *10.4299* | *$10.7 \times 10^{-10}$)* |
| *(Cressida* | *61,767* | *2.4166* | *9.690* | *11.1257* | *$39.5 \times 10^{-10}$)* |
| *(Desdemona* | *62,659* | *2.4515* | *9.620* | *11.3676* | *$20.5 \times 10^{-10}$)* |
| *(Juliet* | *64,358* | *2.5180* | *9.492* | *11.8336* | *$64.1 \times 10^{-10}$)* |
| *(Portia* | *66,097* | *2.5861* | *9.366* | *12.3167* | *$195 \times 10^{-10}$)* |
| 2003U2R ring | ~67,300 | ~2.63 | ~9.28 | ~12.65 | ?? |
| *(Rosiland* | *69,927* | *2.7359* | *9.106* | *13.4030* | *$29.2 \times 10^{-10}$)* |
| *(Cupid* | *74,392* | *2.9106* | *8.828* | *14.708* | *$0.4 \times 10^{-10}$)* |
| *(Belinda* | *75,255* | *2.9444* | *8.777* | *14.9646* | *$41.1 \times 10^{-10}$)* |
| *(Perdita* | *76,416* | *2.9898* | *8.710* | *15.312* | *$1.5 \times 10^{-10}$)* |
| *(Puck* | *86,004* | *3.3649* | *8.210* | *18.2840* | *$333 \times 10^{-10}$)* |
| 2003U1R ring | ~97,700 | ~3.82 | ~7.70 | ~22.15 | ?? |
| *(Mab* | *97,736* | *3.8239* | *7.700* | *22.152* | *$1.0 \times 10^{-10}$)* |
| *(Miranda* | *129,847* | *5.0803* | *6.680* | *33.9235* | *$7{,}590 \times 10^{-10}$)* |

* Mean distance of ring center from the center of the planet for narrow rings, many of which are non-circular (see Table 7.2). In mathematical terms, this is the semi-major axis of the orbits of ring particles located midway between the inner and outer ring boundaries.

in the previous paragraph, *but in the opposite direction*. In other words, the longitude at which an inclined ring crosses the equator of Uranus moves backward around the planet at rates of between 1° and 3° each day.

Optical depth is the term used by astronomers to describe how much of the light incident on a ring (or other target) passes through it. The less starlight that a ring passes, the more optically dense it is. When the incident light is reduced by a factor of $e$ ($=2.718$, the base of natural logarithms), the object through which the light passes is

**Table 7.2.** Uranus ring orbital characteristics.

| Ring region | Mean (km) | Width (km) | Eccentricity | Inclination (deg) | Equiv. depth (km)* |
|---|---|---|---|---|---|
| $\zeta$ (Zeta) ring | 39,600 | ~3,500** | ~0 | ~0 | ~0.25 |
| 6 (Six) ring | 41,837.2 | 1 to 3 | 0.001 01 | 0.062 | .47P, 0.47R |
| 5 (Five) ring | 42,234.8 | 2 to 3 | 0.001 90 | 0.054 | 0.93P, 1.00R |
| 4 (Four) ring | 42,570.9 | 2 to 3 | 0.001 06 | 0.032 | 0.62P, 0.73R |
| $\alpha$ (Alpha) ring | 44,718.4 | 4 to 10 | 0.000 76 | 0.015 | 3.05P, 3.13R |
| $\beta$ (Beta) ring | 45,661.0 | 5 to 11 | 0.000 44 | 0.005 | 1.95P, 2.08R |
| $\eta$ (Eta) ring | 47,175.9 | 1 to 2 | 0.000 00 | 0.001 | 0.62P, 0.48R |
| $\gamma$ (Gamma) ring | 47,626.9 | 1 to 4 | 0.001 09 | 0.000 | ~2.5P, ~4.1R |
| $\delta$ (Delta) ring | 48,300.1 | 3 to 7 | 0.000 04 | 0.001 | 2.34P, ~2.4R |
| $\lambda$ (Lambda) ring | 50,023.9 | 2 to 3 | ~0.0 | ~0.0 | 0.17P, 0.00R |
| $\varepsilon$ (Epsilon) ring | 51,149.3 | 20 to 96 | 0.007 94 | 0.000 | 47.2P, 48.1R |
| 2003U2R ring | ~67,300 | 3,800 | ~0 | ~0 | ~0.02 |
| 2003U1R ring | ~97,700 | 17,000 | ~0 | ~0 | ~0.14 |

* Equivalent depths are from two sources: those labeled 'P' are from stellar occultation measurements by the Voyager 2 photopolarimeter; 'R' values are from radio science occultation. The latter values have been divided by 2 to make them equivalent to the non-coherent photopolarimeter measurements. There are likely variations of equivalent depth around the rings; the measurements here are for single cuts.

** In addition to a uniform dust band of 3,500-km width, there is an inward extension of the Zeta (1986U2R) ring with gradually decreasing brightness to a distance of about 32,600 km [3].

said to have an optical depth ($\tau$) of 1. A reduction by $e^2$ (=7.389) corresponds to optical depth 2. A perfectly opaque object has infinite optical depth; a perfectly transparent object has zero optical depth.

The angle of incident starlight with respect to the plane of Uranus's rings varies with time as Uranus orbits the Sun and (to a lesser degree) as Earth-bound observers of Uranus move in Earth's orbit. If the Uranian rings are more than one ring particle in vertical thickness, the observed optical depth will also vary with viewing angle. If the angle between the viewing direction and a direction perpendicular to the ring plane is $B$, then the normal optical depth, $\tau_n$, can be determined from the observed optical depth, $\tau$, by the relationship

$$\tau_n = \tau \sin B.$$

Two other definitions which are useful in describing the rings of Uranus are equivalent width, $E$, defined as

$$E = W(1 - e^{-\tau}) \sin B$$

and equivalent depth, $A$, defined as

$$A = W\tau_n,$$

where $W$ is the measured radial width of the ring. Note that for rings with very small optical depths, $(1 - e^{-\tau})$ is essentially the same as $\tau$, and $E \approx A$. For larger optical depths, $E$ is always less than $A$.

Using the definitions for equivalent width and depth to describe Uranus ring observations reveals something of their nature. If the rings were monolayers (only a single particle thick), there should be no variations of $E$ around the rings. The four eccentric rings that show systematic variations of width with distance from the planet (Alpha, Beta, Delta, and Epsilon rings) all show variations of $E$ with ring azimuth. These rings, and by analogy all of the Uranian rings, are not monolayers, but are many particles thick. In spite of that conclusion, recent observations of the Uranus rings from the Hubble Space Telescope at low tilt angles (the Uranus rings are edge-on to the Sun and Earth in 2007) seem to show that the narrow rings of Uranus may be the only true monolayered rings in the solar system [4].

Equivalent depth, $A$, provides a measure of the amount of material in a ring, and might be expected to be relatively constant around a ring. The ground-based observations were not conclusive on this point. Values of $A$ around any ring tend toward constancy, but the measured variations are larger than the estimated errors in the measurements. These variations in $A$ could be due either to clumpiness (azimuthal non-uniformity in the numbers of ring particles) or to substantial radial variations in optical depth across a given ring. Radial variations could invalidate the assumption that equivalent depth, defined by multiplying the measured width by an *average* optical depth, is representative of the number of ring particles. The ground-based observations of the rings suggest that both clumpiness and radial non-uniformity contribute to the noted azimuthal variations in equivalent depth.

There is probably a close relationship between the equivalent depth of a ring and its total mass. Ring masses are sometimes estimated on the basis of their average optical depths or their equivalent depths. A more direct method of estimating ring mass is by studying density waves created in a ring by nearby satellites. Unfortunately, only one such wave has been identified in any of the Uranian rings; that identification was for a density wave in the Delta ring detected in Voyager photopolarimeter data [5]. A rough estimate of the *surface density* in the Delta ring was made from measurements of the observed density wave; it yielded values of between 5 and 10 grams per square centimeter. The density wave, if it truly exists, must be due to an as-yet-undiscovered satellite near the Delta ring. The process of deriving surface masses from measurements of density waves requires the assumption of an extended, nearly homogenous disk, and occultation measurements of the Delta ring show that it is neither extended nor homogenous. Masses of the Uranian rings therefore remain essentially unknown, as indicated in the last column of Table 7.1.

## 7.3 PHYSICAL PROPERTIES OF THE RING PARTICLES

The Uranus ring data from Voyager 2 span the electromagnetic spectrum from ultraviolet to radio wavelengths. The data also cover a range of viewing and illumination geometries. Intercomparison of the data sets provides information relevant to studies of the physical properties of individual ring particles, even though the particles themselves are too small to be resolved in the data sets. Voyager results are also descriptive of the environment experienced by the ring particles and are therefore useful for theoretical studies of ring evolution. Combined with the Earth-based results they provide a relatively detailed database from which to extract information on the physical properties of the rings of Uranus.

### 7.3.1 Ring-particle size distribution

Particle sizes in Jupiter's Main ring are clustered near a radius of 1 micrometer. Those in the Halo ring are smaller still. Particles in the Saturn ring range from micrometer-sized to house-sized or larger. Much of the information comes from the observed differences in transparency of the rings at different wavelengths. In radio and stellar occultation experiments, the ring particles tend to absorb radiation at wavelengths comparable with or smaller than their diameters and to pass without significant absorption radiation at longer wavelengths.

The nine Uranian rings first seen from Earth seem to have nearly identical optical depths at all measuring wavelengths (ultraviolet to radio). The implications are clear: the vast majority of the ring particles must have sizes comparable with or larger than the longest observing wavelength (13 cm). A single long-exposure image of Uranus's rings at the highest possible phase angle (172.5°) is shown in Figure 7.1. In the image, taken from an article by Burns [6], the rings are seen to be almost continuous, but with a large amount of radial variation. The fact that all the rings seen at low phase angle (top half of image) are visible in the high-phase-angle image is an indication that a small fraction of the particle population in each of the narrow rings is composed of particles with radii of approximately 1 micrometer. In addition, there is a sprinkling of micrometer-sized dust between the narrow rings; the source of the radial structure in the dust bands, much like that of the narrow rings themselves, remains unexplained.

The brightest of the dust rings is at the radial distance of the Lambda ring, which is very faint in backscattering and was not seen in radio science occultation data. Combined, these data indicate that the Lambda ring is dominated by particles in the size range of 1 micrometer radius or smaller. The equivalent depth of the Lambda ring measured by the ultraviolet spectrometer (wavelength 0.31 micrometer) was 0.31 km; that measured by the photopolarimeter (wavelength 0.27 micrometer) was only 0.13 km. These data suggest that more than 60% of the particles in the Lambda ring are less than 0.3 micrometers in radius.

The three broad rings of Uranus have not been detected in Earth-based stellar occultation measurements. All three have been seen in backscattered near-infrared light from Earth. They are probably dominated by micrometer-sized or smaller particles, but, like the Lambda ring, they must have a tiny fraction of larger particles which enable them to be seen at low phase angles, especially when the tilt angle of the Uranus rings is small.

Voyager radio occultation data are of high enough quality to identify faint companions of two of the rings of Uranus. The Eta ring has a faint companion ring at its outer edge; the Delta ring has a similar companion ring at its inner edge. These companions were seen in both ingress and egress measurements with similar size and radio optical depth in each case, even though the two cuts were separated by about 149° in ring longitude [7]. The Delta and Eta ring companions have also been seen in

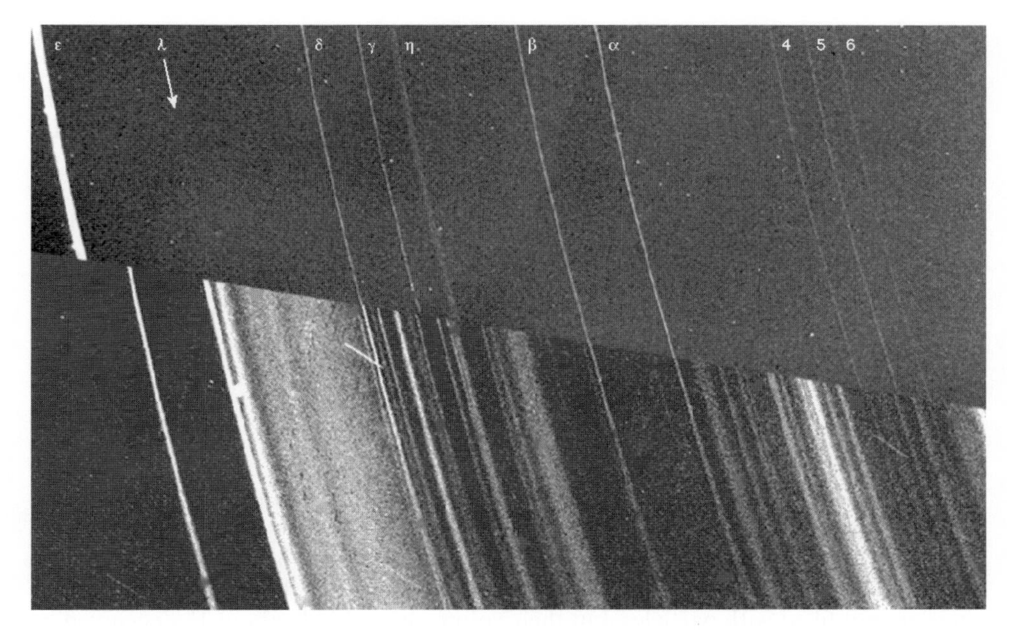


**Figure 7.1.** Two views of the Uranus rings as seen by Voyager 2 are compared in this composite image of the ring system at low (upper image) and high (lower image) solar phase angles. Most of the structure seen in the lower image is micrometer-sized ring particles which are visible in no other images. All of the rings are seen at low phase angle and also present at high phase angle, although some are shifted in apparent radial position, either by ring eccentricity (as in the Epsilon ring) or by a combination of ring inclination and eccentricity (Four, Five, and Six rings). Image from Burns [6].

Earth-based stellar occultations [8]. Their optical depths appear to be constant with both time and wavelength, implying again that these are not dust rings, but are instead dominated by particles larger than a few centimeters.

### 7.3.2 Ring-particle composition

No high-resolution ultraviolet, visible, or infrared spectra of the rings of Uranus were obtained by Voyager 2. The rings were too dark, too narrow, and too optically thin to permit the collection of such data. Color images shuttered at different wavelengths in the visible range of the spectrum could potentially provide some composition information. The rings of Uranus were uniformly dark at all colors. The only cosmically abundant material which matches both the low reflectivity and the uniformly gray color of the Uranus rings is carbon. The five largest satellites of Uranus also have neutral colors, but their reflection spectra indicate that there is water ice present on their surfaces. No indication of water ice is seen in Earth-based spectra of the rings, although the two recently discovered outer rings of Uranus (2003U1R and 2003U2R) appear to be blue and red, respectively [9]. The blueness of 2003U1R and the presence of a satellite (Mab) near its peak brightness bring to mind Saturn's E ring, the only other known ring with a bluish hue. By analogy, Mab could be the source of the ring's material, which is almost certainly dominated by tiny particles of water ice. The redness of 2003U2R is reminiscent of Saturn's G ring and may be indicative of a silicate composition.

The source of the carbon in the main rings of Uranus is very possibly the decomposition of methane ($CH_4$) ice through bombardment by energetic protons [10]. Methane is known to be abundant in the outer solar system, and energetic proton fluxes are sufficiently high to blacken most exposed methane in the inner magnetosphere. It is also possible that part or all of the carbon is in the form of originally elemental carbon from the breakup of carbonaceous meteoroids or asteroids captured into orbit around Uranus.

### 7.3.3 Ring-particle reflectivity

The low reflectivity of the rings of Uranus against the background sky is evident both from Voyager 2 and from Earth-based observations. Reflectivity of the Epsilon ring is estimated to be between 1.0% and 1.8%, making it one of the darkest natural structures known [11]. The other narrow rings have similarly low reflectivity. The rings are not filled with particles, but their volume is probably about 1% occupied by particles. A flat sheet composed of the same material as the ring particles might have a reflectivity of about 3.2%. For comparison's sake, the A and B rings of Saturn have reflectivity of about 60%, and the darker C ring and Cassini Division reflect between 20% and 30% of the sunlight incident upon them [12].

## 7.4 RING INTERACTIONS WITH SATELLITES

One of the ring science objectives of the Voyager 2 encounter was to study interactions between ring particles and the shepherding satellites most scientists expected to find between the rings. Such interactions are responsible for some of the major features of Saturn's rings, such as the outer edges of the A and B rings and a large number of less prominent features spread throughout the rings. Small satellites are also thought to be the source of new ring material, either by tidal breakup as these satellites approach the planet too closely and are broken apart by unmatched gravitational forces, or in less catastrophic manner by countless micrometeoroid impacts with their surfaces. Undiscovered satellites with diameters of 5 to 10 kilometers may still await discovery near the orbits of the Uranus rings. Even smaller satellites could serve as sources for the material in the present rings.

As matters presently stand, only the Epsilon ring seems to have satellites in the proper positions to possibly confine ring material within the variable width of the ring. Observations seem to show that the ring is optically denser near its inner and outer edges, typical of rings that are gravitationally confined. Ophelia, orbiting farther from Uranus than the Epsilon ring, completes 13 orbits of Uranus for every 14 orbits of particles at the outer edge of the Epsilon ring. Similarly, Cordelia orbits the planet 25

**Figure 7.2.** Computer processing of one-dimensional radio science occultation data were used to produce this two-dimensional view of the Epsilon ring showing many internal details, including very abrupt termination of ring material at inner and outer boundaries. Also apparent in the image are the abrupt rise in brightness near the center of the ring and the higher concentration of ring material in the outer half of the ring. (Courtesy G. L. Tyler)

times for every 24 orbits of ring particles at the inner edge of the Epsilon ring. The repetitive geometric alignment of ring particles and satellites can result in a slight retardation in the orbital speeds of particles at the outer ring edge and acceleration of particles near the inner edge. Those combined gravitational nudges may be the confining mechanism for particles in the Epsilon ring, which, due to mutual collisions between ring particles, would otherwise tend to spread radially both inward and outward. A 47:49 *orbital resonance* with Cordelia which falls near the center of the Epsilon ring may additionally account for that ring's relative minimum in optical depth near its center and for the larger number of particles confined to the outer half of the ring (see Figure 7.2).

The interaction is complicated somewhat by the fact that the orbit of Ophelia is slightly elliptical (eccentricity of 0.010). The average eccentricity of the Epsilon ring is intermediate between the negligible eccentricity of Cordelia's orbit and the larger eccentricity of Ophelia's orbit; this may be related to the fact that the Epsilon ring is widest at its most distant point from the planet and narrowest at its closest approach to the planet.

As successful as gravitational resonance theory with known satellites may be in explaining the major features of the Epsilon ring, it has thus far been singularly unsuccessful in explaining the finer details of the Epsilon ring or in explaining most of the boundaries of the other narrow rings of Uranus. The 23:22 resonance of Cordelia falls very close to the sharp outer edge of the Delta ring (but is not the cause of the Delta ring density wave mentioned above), and the 6:5 resonance of Ophelia is close to the outer edge of the Gamma ring [13]. Aside from these perhaps fortuitous matches, the satellites (or other mechanisms) responsible for ring confinement are unknown.

## 7.5 DUST ENVIRONMENT AND POSSIBLE MAGNETIC FIELD INTERACTIONS

Voyager 2 crossed the Uranus equatorial plane near a distance of 115,400 km from the center of the planet. That is about 14,400 km inside the orbit of Miranda and well exterior to any of the rings of Uranus. Two instruments aboard Voyager 2 (plasma wave and planetary radio astronomy instrumentation) were capable of detecting dust impacting the spacecraft. These instruments detected a maximum of 30 to 50 impacts per second very near the Uranus equator and extending more than 1,000 km above and below the equatorial plane [14]. From the measurements, estimated to be due to particles in the size range of a few micrometers or so, a maximum spatial density of about 1 dust particle per 1,000 cubic meters was calculated, dropping by a factor of 2 for every 100 to 150 km from the Uranus equator.

One possible reason for the vertical spreading of this dust ring may be electrical charging of the dust particles by constant bombardment from positively charged ions or electrons. Because of the high tilt of the magnetic field of Uranus, ring and dust particles are blasted by these plasma particles twice every rotation of Uranus. The sole exception to that general rule is at the radial distance from Uranus where the rotation rate of the planet is equal to the orbital period of ring particles, which occurs near a radius of about 82,000 km, just interior to the orbit of Puck and about halfway between the two recently discovered rings of Uranus. The strongest perturbing forces tending to move particles out of the Uranus equatorial plane would be expected to occur where ring particles complete two orbits of the planet per single rotation of Uranus, and where ring particles complete one orbit of the planet per two rotations of Uranus. The former occurs just exterior to the Epsilon ring, possibly providing a mechanism for removing small particles from that ring. The latter occurs just beyond the orbit of Miranda, thereby providing a possible reason for the large vertical extent of the dust band observed by the two Voyager instrument packages.

## 7.6 NOTES AND REFERENCES

[1] The data in Table 7.1 are taken from tables 11.1 and 11.2 in Miner, E. D., 1997, *Uranus: The Planet, Rings and Satellites*, Wiley-Praxis Series in Astronomy and Astrophysics, Chichester, England; the data for the newly discovered rings is from Planetary Rings Node at *http://pds-rings.seti.org/uranus/uranus_tables.html.*

[2] The data in Table 7.2 are from the same sources as those in Table 7.1 [1].

[3] De Pater, I., Gibbard, S. G., Hammel, H. B., 2006, "Evolution of the dusty rings of Uranus", *Icarus* **180**, 186–200. The nomenclature of the innermost ring of Uranus as the $\zeta$ (Zeta) ring was first adopted by these authors in place of the temporary designation as 1986U2R; it is assumed that this name will be approved by the International Astronomical Union.

[4] De Pater, I., Gibbard, S. G., Hammel, H. B., 2004, "Uranus's ring 1986U2R detected with Keck AO at 2.2 microns", *Bulletin of the American Astronomical Society* **36**, 1110 (abstract only), and associated press release from University of California at Berkeley, 10 November 2004.

[5] Horn, L. J., Yanamandra-Fisher, P. A., Esposito, L. W., Lane, A. L., 1988, "Physical properties of the Uranian $\delta$ ring from a possible density wave", *Icarus* **76**, 485–492.

[6] Burns, J. A., 1999, "Chapter 16: Planetary Rings", in *The New Solar System* (4th Edition), edited by Beatty, Peterson, and Chaikin, pp. 221–240. The relevant figure is figure 14 on p. 230.

[7] Gresh, D. L., Marouf, E. A., Tyler, G. L., Rosen, P. A., Simpson, R. A., 1989, "Voyager radio occultation by Uranus' rings. I. Observational results", *Icarus* **77**, 131–168.

[8] Elliot, J. L., Nicholson, P. D., 1984, "The rings of Uranus", in *Planetary Rings*, edited by Greenberg and Brahic, pp. 25–72.

[9] De Pater, I., Hammel, H. B., Gibbard, S. G., Showalter, M. R., 2006, "New dust belts of Uranus: One ring, two ring, red ring, blue ring", *Science* **312**, 92–94.

[10] Cheng, A. F., Haff, P. K., Johnson, R. E., Lanzerotti, L. J., 1986, "Interactions of planetary magnetospheres with icy satellite surfaces", in *Satellites*, edited by Burns and Matthews, pp. 403–436.

[11] Esposito, L. W., Brahic, A., Burns, J. A., Marouf, E. A., 1991, "Particle properties and processes in Uranus' rings", in *Uranus*, edited by Bergstralh, Miner, and Matthews, pp. 410–465.

[12] Cuzzi, J. N., Lissauer, J. L., Esposito, L. W., Holberg, J. B., Marouf, E. A., Tyler, G. L., Boischot, A., 1984, "Saturn's rings: Properties and processes", in *Planetary Rings*, edited by Greenberg and Brahic, pp. 73–199.

[13] Porco, C. C., Goldreich, P., 1987, "Shepherding of the Uranian rings. I. Kinematics", *Astronomical Journal* **93**, 724–729.

[14] Esposito, L. W., Brahic, A., Burns, J. A., Marouf, E. A., 1990, "Particle properties and processes in Uranus' rings", in *Uranus*, edited by Bergstralh, Miner, and Matthews, pp. 410–465.

## 7.7 BIBLIOGRAPHY

Elliot, J. L., Nicholson, P. D., 1984, "The Rings of Uranus", in *Planetary Rings*, edited by Greenberg and Brahic, pp. 25–72.

Esposito, L. W., Brahic, A., Burns, J. A., Marouf, E. A., 1990, "Particle properties and processes in Uranus' rings", in *Uranus*, edited by Bergstralh, Miner, and Matthews, pp. 410–465.

French, R. G., Nicholson, P. D., Porco, C. C., Marouf, E. A., 1990, "Dynamics and structure of the Uranian rings", in *Uranus*, edited by Bergstralh, Miner, and Matthews, pp. 327–409.

Miner, E. D., 1997, *Uranus: The Planet, Rings and Satellites* (2nd Edition), Wiley-Praxis Series in Astronomy and Astrophysics, 360 pp.

## 7.8 PICTURES AND DIAGRAMS

Figure 7.1 *http://photojournal.jpl.nasa.gov/jpeg/PIA00035.jpg* and *http://photojournal.jpl.nasa.gov/jpeg/PIA00142.jpg*

Figure 7.2 Miner fig. 11.6 (courtesy G. L. Tyler). Miner, E. D., 1997, *Uranus: The Planet, Rings and Satellites* (2nd Edition), Wiley-Praxis Series in Astronomy and Astrophysics, 360 pp.

# 8

# Present knowledge of the Neptune ring system

## 8.1 INTRODUCTION

In Chapter 5, we discussed the discovery and early observations (including some of the Voyager observations) of the Neptune ring system, last of the ring systems to be discovered around a giant planet in the solar system. All of the four giant planets of the solar system have rings, and none of the terrestrial planets possess such rings. That difference must have had its origin in the processes and subsequent evolution that shaped the giant planets.

Voyager 2 is the only spacecraft that has visited Neptune. The spacecraft's closest approach to Neptune occurred at 4 a.m. Greenwich Mean Time on August 25, 1989. The only close-range data collected on Neptune's ring system were collected over a 2-week period surrounding that date of closest approach; Voyager 2 data are also the sole available data for phase angles larger than a few degrees. Only recently has Earth-based telescopic equipment achieved the sensitivity and resolution to begin additional low-phase-angle observations of Neptune's tenuous rings. Hence, the most comprehensive treatise to date on the Neptune ring system is a compilation of pre-Voyager and Voyager observations and their interpretation included as a chapter by Porco *et al.* [1] in the University of Arizona Press text *Neptune and Triton*, edited by D. P. Cruikshank. A somewhat less formal and much shorter summary is given in chapters 3 and 11 of the book *Neptune: The Planet, Rings and Satellites*, written by two of the authors of this book [2].

Nearly 800 images of the Neptune ring system were returned by Voyager 2. A handful of observations were also carried out by other investigations aboard Voyager 2. These included a stellar occultation measured by the photopolarimeter and the ultraviolet spectrometer, in which the light received from the star sigma Sagittarii was carefully monitored by Voyager 2 as the spacecraft motion caused Neptune's rings to pass across the star, partially blocking the starlight. Voyager's very precise X-band and S-band radio signals were also monitored on Earth as the spacecraft passed behind the rings in what is known as a radio occultation experiment. Dust and plasma near the spacecraft were also measured by the Voyager plasma wave, planetary radio astronomy, cosmic ray, and low-energy charged particle instruments. All of these data assisted ring scientists to better understand the nature of the Neptune rings. However, it is clear that there remain many questions that may not be answered without additional data from a future Neptune-orbiting spacecraft similar to the Cassini Orbiter at Saturn or by far more sophisticated Earth-based observations than are possible today.

## 8.2 RING RADIAL CHARACTERISTICS

Neptune is thirty times as far from the Sun as the Earth is. That translates to sunlight that is a scant 1/900th as bright as sunlight at Earth, or something akin to late twilight on Earth. Combine with that the fact that Neptune's rings are inherently dark (as well as being optically thin) and the problem of imaging Neptune's rings becomes something like trying to image pieces of coal when the sole illumination is a full Moon. Because of the dust content of the rings, they become somewhat easier to see when back-lighted, so some of the best images of Neptune's rings come from phase angles (the angle between the Sun and the observer as seen from the target) of about 135°. Most of the images of the Neptune rings were shuttered either during approach to the planet, where the phase angle was about 15°, or during departure, where the phase angle was about 135°. Only a few images obtained about 6.5 hr before closest approach (phase angle ~8°) and 37 to 77 min after closest approach (phase angle ~155°) were appreciably different from the approach and departure phase angles.

A table of the radial structure of the rings of Neptune was given earlier in Chapter 5 (Table 5.3). There are three continuous narrow rings (Adams, Le Verrier, and Arago), one faint and possibly intermittent narrow ring (which shares it orbit with the satellite Galatea), and two broad rings (Lassell and Galle). A radial scan of the measured brightness of the rings at 134° phase angle is shown in Figure 8.1 [3].

The radio science occultation experiment, which had yielded such a rich store of data for the Saturn and Uranus ring systems (no radio occultation of the Jupiter ring was attempted), yielded very little for the Neptune ring system. Tyler *et al.* [4] reported detecting no Neptune ring material down to the noise level of the radio signal, which corresponded to about 1% of the received signal strength for a radial resolution of 2 km. The team had three caveats on their non-detection: (a) only the Adams ring was probed on both the immerging and the emerging sides of Neptune and the Galle ring was not probed at all, (b) neither side probed the longitudes in the Adams ring that contained the denser ring arcs (see Section 8.3), (c) the radio data were affected by passage through Neptune's ionosphere, thus perhaps masking a weak ring signature.

The stellar occultation measurements of sigma Sagittarii occurred as Voyager 2 was inbound toward the planet, about 5 hr from closest approach. The occultation covered a range of radial distances from Neptune of 42,414 to 76,056 km. The lower end of this range is unfortunately at the outer edge of the Galle ring, but the upper end of the range is well outside the Adams ring. The Adams ring was detected by both the

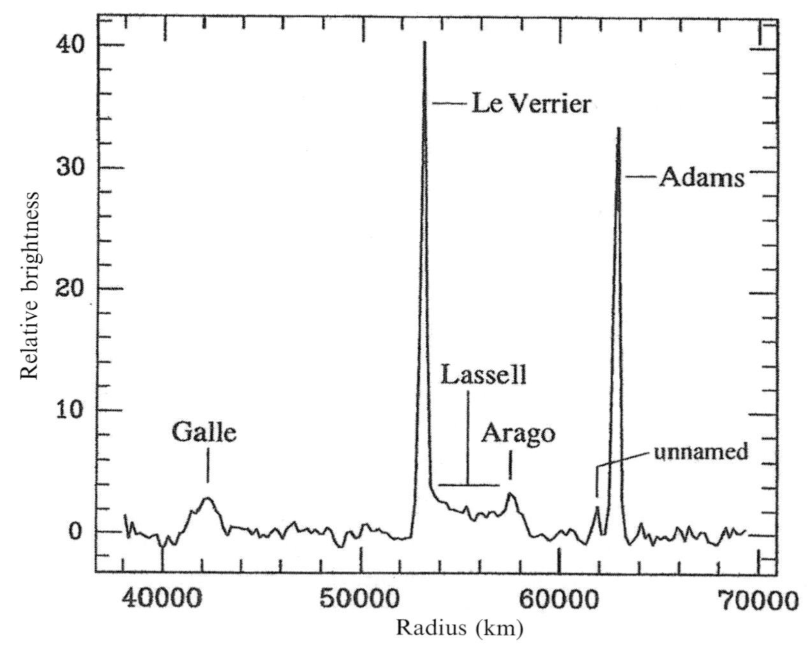


**Figure 8.1.** This plot of relative brightness versus radial distance was produced by Mark Showalter by radially scanning a wide-angle image (FDS 11412.51) of the rings taken at a phase angle of 135°. The data are azimuthally averaged to reduce image noise. The six rings of Neptune are clearly seen in the plot, which appears as fig. 3 of Porco *et al.* [1].

photopolarimeter (with an effective wavelength of 0.26 micrometers) [5] and the ultraviolet spectrometer (with an effective wavelength of 0.11 micrometers) [6]. Both wavelengths are well below the range of visible light wavelengths in the far ultraviolet. The value of the radial distance depends somewhat on the inclination of the Adams ring, which is not well determined, but is close to 62,900 km, in reasonable agreement with the value of 62,932 given in Table 5.3.

At this point, let us pause for a moment to remind our reader of the concept of *equivalent depth* introduced in Chapter 7 in the discussion of the variable-width Uranus rings. Equivalent depth is basically the optical depth (mathematically corrected to vertical viewing) times the physical width (in kilometers) and is a measure of the total material in a cross-section of a ring. Optical depth (= optical thickness) is a measure of the amount of light absorbed in passage through a ring and is indicated as the natural logarithm of the ratio of the intensity of the incident light to that of the emerging light. In simple terms, a ring (or other semi-transparent sheet of material) is said to have an optical depth of 1 if the incident light is reduced by a factor of $e$ (= 2.718). The optical depth is 2 if the reduction is a factor of $e^2$ (= 7.389), and so forth. Vertical (or normalized) optical depth is the measured optical depth multiplied by the trigonometric sine of the viewing angle. (The viewing angle is the angle between the observing direction and the perpendicular to the ring plane.) Normalized optical depth is therefore an approximation of the optical depth of the rings for vertical illumination and viewing. In mathematical terms,

$$I = I_0 e^{-\tau} \quad [\text{or alternatively expressed as } \tau = \ln(I_0/I)],$$

where $I_0$ is the incident light intensity, $I$ is the emerging light intensity, $e$ is the base of natural logarithms, ln is the natural logarithm, and $\tau$ is the optical depth,

$$\tau_n = \tau \sin B,$$

where $B$ is the viewing angle just described and $\tau_n$ is the normalized optical depth, and

$$A = W\tau_n,$$

where $A$ is the equivalent depth and $W$ is the physical width. For the rings of Uranus, equivalent depth tends to be relatively constant around the narrow rings, even when their physical widths vary. Equivalent depth is, in a sense, a measure of the amount of ring material in a cross-section of the ring.

Now back to our discussion of the stellar occultation measurements of the Neptune rings by Voyager 2. To reduce the noise in the data, the photopolarimeter, with a sampling interval of 1.5 km, was smoothed to an effective resolution of 5 km. The equivalent depth of the Adams ring as measured by the Voyager photopolarimeter [7] was determined to be $0.77 \pm 0.13$ km. The ultraviolet spectrometer, with a radial resolution of 2.3 km, measured an Adams ring equivalent depth [8] of $0.66 \pm 0.12$ km. The two numbers are statistically identical, an indication that there are few ring particles in a size range near 0.1-micrometer radius in that part of the Adams ring, which fortuitously corresponds to the leading edge of the Liberté arc within the Adams ring. The Liberté arc is much brighter than those portions of the Adams ring sampled by the radio occultation experiment (60° and 90° from the arc region); therefore, no conclusions can be drawn about the relative numbers of larger ring particles in the Adams ring.

The ultraviolet spectrometer detected no Neptune ring features other than the Adams ring. A statistical analysis of the photopolarimeter data [9] indicated that the Adams ring occultation was the only unambiguous non-random event detected during the occultation experiment. However, there is a slight dip in the starlight intensity near the radial distance of the Le Verrier ring, which may indicate a near detection of that ring. Otherwise, the most useful data on the radial positions of the six Neptune rings listed in Table 5.3 are those from the Voyager imaging data.

Beyond the rings seen by the imaging system, Voyager detected dust particles near the ring plane both inbound and outbound. The detections were made by the plasma wave [10] and planetary radio astronomy [11] investigations. The inbound equatorial crossing of Neptune was at a radial distance of 85,290 km and the outbound was at a radial distance of 103,950 km [12]. On the inbound leg, both investigations found a maximum in the numbers of dust particles at a radial distance of 85,400 km and a vertical height above the ring plane of +146 to +160 km. The numbers of particles dropped off smoothly above and below that point, reaching half the maximum number density at a distance of about $\pm 140$ km. The two investigations differed slightly in their derived numbers for the outbound ring crossing. The plasma wave investigation found the maximum numbers near a radial distance of 104,000 km and near a vertical distance of −948 km below Neptune's equatorial plane; the vertical width of the dust distribution was more than ±500 km at half maximum. The planetary radio astronomy investigation found a maximum at a slightly larger radial distance of 105,500 km and a little closer to Neptune's equator (−700 km); they found a much narrower dust distribution with a thickness of about ±115 km.

## 8.3 RING AZIMUTHAL VARIATIONS

No significant azimuthal inhomogeneity has been noted in Neptune's inner four rings. The unnamed ring that shares its orbit with the satellite Galatea is not visible in most Voyager images of the rings; for those in which it is discernible, it appears to be discontinuous, but the existing data are insufficient to quantify any azimuthal variability. The only ring for which azimuthal variability is clearly and measurably present is the Adams ring.

The primary azimuthal structure noted in the Adams ring is associated with the ring arcs. A radially averaged longitudinal scan of a Voyager 2 wide-angle image that includes all the arcs is shown in Figure 8.2, taken from Porco *et al.* [13]. The arcs, from leading to trailing, are Courage, Liberté, Egalité 1, Egalité 2, and Fraternité. Their relative positions and respective lengths were given in Table 5.4. Internal brightness variations within the ring arcs, clearly evident from Figure 8.2, are real and exist in each of the arcs. These variations are likely due to clumpiness within the arcs. They may be the result of accumulation of dust-sized particles around larger than average bodies within the arcs. The larger bodies in turn may be the source bodies for the smaller material in the arcs [14].

In addition to the fine-scale brightness variations within the ring arcs, the non-arc regions of the Adams ring seem to vary by about a factor of 3; the background ring is brightest near and between the ring arcs and faintest well away from the longitude of the arcs [15].

Recent observations of the Adams ring arcs from the Keck Telescope, utilizing its adaptive optics capability, show that there have been substantial changes in the ring arcs since the Voyager 2 encounter in 1989 [16]. The trailing arc, Fraternité, seems to be the only well-behaved arc within the ring. It continues to be essentially unchanged in appearance and to circle Neptune at the same rate measured during the Voyager era (820.1118 deg/day). Egalité 1 and Egalité 2 appear to have reversed in relative intensity, perhaps the result of material migrating between the two resonance sites. Courage appears to be approximately 8° ahead of its prior position relative to Fraternité, perhaps an indication that it has shifted one resonance site ahead of its prior position. In earlier data, Liberté seemed to be migrating between resonance sites; in the latest data it has all but disappeared.

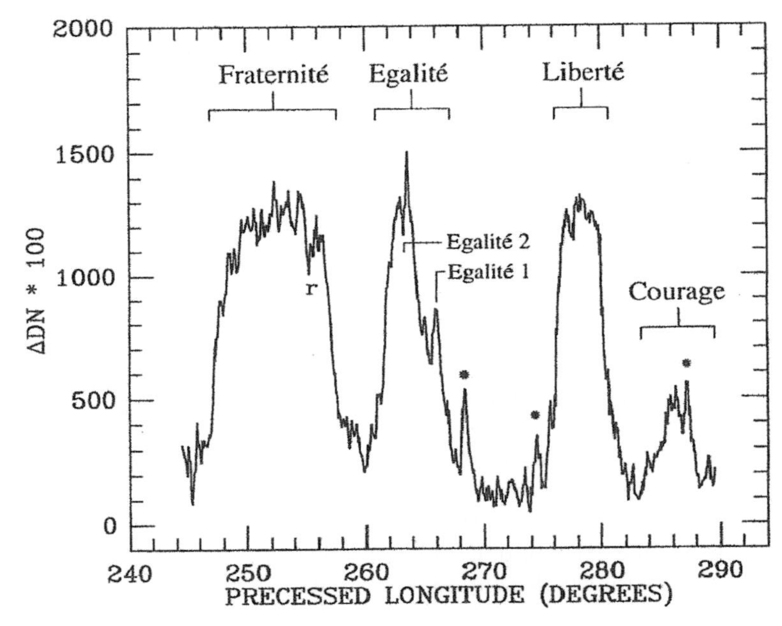


**Figure 8.2.** In a wide-angle image obtained 20.5 hr after Voyager 2's closest approach to Neptune, all the Adams ring arcs were captured. A radially averaged azimuthal scan of the Adams ring depicts the relative sizes and positions of the five ring arcs. The longitude system is one for which the rings are fixed as of August 18, 1989. Adjustment to that date was accomplished by precession of the observed position backwards at a rate of 820.1185 deg/day, the observed rotational rate of the A ring arcs. The three asterisks in the figure indicate the positions of background stars. The "r" in the Fraternité arc shows the position of an incompletely removed black reseau mark from the camera; reseau marks are used for geometric reconstruction of Voyager images.

## 8.4 SMALL SATELLITES NEAR THE RINGS

Four known satellites orbit near or within the Neptune ring system (see Table 8.1). These include Naiad, Thalassa, Despina, and Galatea. Naiad, Thalassa, and Despina are all between the Galle and Le Verrier rings. Galatea seems to circle Neptune precisely 43 times for every 42 circuits of particles in the Adams ring. In scientific terms, that means that the Adams ring and its ring arcs are located at a 42:43 *outer Lindblad resonance* [17] of Galatea. Galatea may therefore be partly responsible for the narrowness of the Adams ring. It may also play a role in the azimuthal confinement of the ring arcs. Over time, its gravitational influence will create 42 nodes in the Adams ring which may inhibit the azimuthal motion of ring particles past those nodes, which would be separated by 8.57° of ring longitude. Six nodes would span a longitudinal range of about 42.85°, very nearly the same as the longitude span covered by the five known arcs.

**Table 8.1.** Neptune ring dimensions ($1R_N = 24{,}674$ km, $1M_N = 10.246 \times 10^{25}$ kg).

| Ring region | Boundary (km) | Boundary ($R_N$) | Velocity, (km/s) | Period (hr) | Mass ($M_N$) |
|---|---|---|---|---|---|
| Galle ring | ~40,900 | 1.66 | 12.94 | 5.518 | ?? |
| | ~42,900 | 1.74 | 12.63 | 5.927 | |
| *(Naiad* | *48,227* | *1.95* | *11.914* | *7.065* | *$2 \times 10^{-9}$)* |
| *(Thalassa* | *50,075* | *2.02* | *11.690* | *7.476* | *$4 \times 10^{-9}$)* |
| *(Despina* | *52,526* | *2.12* | *11.414* | *8.032* | *$20 \times 10^{-9}$)* |
| Le Verrier ring | $53{,}200 \pm 20$ | 2.15 | 11.35 | 8.187 | ?? |
| Lassell ring | 53,200 | 2.15 | 11.35 | 8.187 | ?? |
| | 57,200 | 2.31 | 10.92 | 9.129 | |
| Arago ring | 57,200 | 2.31 | 10.92 | 9.129 | ?? |
| Unnamed ring | ~61,950 | 2.50 | 10.50 | 10.290 | ?? |
| *(Galatea* | *61,953* | *2.50* | *10.508* | *10.290* | *$37 \times 10^{-9}$)* |
| Adams ring | $62{,}932 \pm 2$ | 2.54 | 10.42 | 10.535 | ?? |
| *(Larissa* | *73,548* | *2.97* | *9.643* | *13.312* | *$48 \times 10^{-9}$)* |

*Note*: Widths of the narrow rings (in km) are: Le Verrier ~110; Arago and Unnamed <100; Adams 15–50.

Despina is relatively close to the Le Verrier ring, but its gravitational force seems to have no effect on that ring. The center of the ring is only a few kilometers from the position of a potential 53:52 outer Lindblad resonance, but the 54:53 resonance is only 13.1 km closer to the planet, and the Le Verrier ring has a width of about 110 km. Furthermore, the absence of azimuthal inhomogeneity or internal radial structure would seem to rule out substantial gravitational effects from Despina. If the Le Verrier ring's width is affected by a Lindblad resonance, it does not appear to be that of Despina.

Galatea seems also to be subject to bombardment by meteoroids that have blasted material from its surface. That is probably the source of the tenuous and possibly intermittent ring that shares its orbit. Little more is known about this tenuous ring that has yet to garner an official name from the International Astronomical Union's Nomenclature Commission.

## 8.5 POSSIBLE ELECTROMAGNETIC EFFECTS

The presence of an extended disk of dust well beyond the visible ring system was discussed in Section 8.3. The plasma wave and planetary radio astronomy investigations also detected dust particles at all latitudes, albeit at levels that were several orders of magnitude smaller than observed near the equator. Neptune's magnetic equator is highly tilted (by 47°!) with respect to Neptune's rotation equator, so if tiny dust particles in the extended dust disk become electrically charged, they can perhaps be rapidly moved out of the equatorial plane near Lorentz resonance positions, as discussed for the Halo ring of Jupiter in Chapter 6. Lorentz resonances occur at radial distances where the orbital period of the dust particles is a simple integer ratio of the rotation period of Neptune. Neptune's rotation period is $16.108 \pm 0.006$ hr [18]. By inspection of Table 8.1, it is clear that the co-rotation radius (where the natural orbital period is equal to the rotation period of Neptune) is well outside the rings. It falls at a radial distance of 3.22 $R_N$ (79,740 km) from Neptune's center, between the orbits of Larissa and Proteus. The 3:2 Lorentz resonance radial distance is outside the Adams ring and the 2:1 resonance is interior to the Le Verrier ring. Any electrically charged sub-micrometer-sized particles that wander planetward at those two radial distances will quickly be perturbed to high inclinations, possibly ending up as a part of the high-latitude dust population observed.

Ring particles in the main rings and the dust disk, except at the co-rotation radius near 79,740 km, will have trapped radiation in the Neptune magnetosphere swept back and forth across them. The resultant collisions are expected to remove some of that plasma from the magnetosphere, resulting in reduced plasma levels above and below the rings. However, the passage of Voyager 2 over the pole of Neptune did not take the spacecraft through those portions of the Neptune magnetosphere that might enable such depletions to be measured directly.

## 8.6 PHYSICAL PROPERTIES OF THE RINGS

The absence of a detection of the Adams ring (away from the arc region) implies that the material in that portion of the ring is much smaller than a few centimeters in radius. This is in sharp contrast to similar measurements of the rings of Saturn and Uranus, where dust-sized particles apparently constitute 1% or less of particle population. The brightness of the six Neptune rings at high phase angle in Voyager 2 images leads to a similar conclusion: that a high percentage of the particles in the rings must be micrometer-sized. Similarly, from the greater contrast between the arcs in the Adams ring and particles in the remainder of the ring in high-phase images, the ring arcs have an even higher percentage of micrometer-sized particles than the rest of the Adams ring. The Le Verrier ring particles are similar to those of the non-arc regions of the Adams ring.

Looking at particle sizes in the broader Galle and Lassell rings, the Galle ring also appears to have high dust content. The Lassell ring, of comparable optical depth, appears significantly less dusty (although still more than ten times as dusty as the main rings of Saturn or Uranus).

One characteristic common to the rings of Uranus and Neptune is their low reflectivity. Both are characterized by particles that reflect less than 5% of the light incident on their surfaces. This low reflectivity is certainly uncharacteristic of water ice particles. Perhaps both ring systems have particles coated with black elemental carbon, possibly produced by the bombardment of methane by magnetospheric plasma.

The masses of the rings of Neptune (listed as '??' in Table 8.1) were not determined, but they are estimated to be 10,000 times less massive than the rings of Uranus.

If dedicated searches for such rings had not been conducted, either from Earth or from Voyager 2, they might never have been seen in images taken for other purposes. Neptune has more satellites within its ring system (Naiad, Thalassa, Despina, and Galatea) than Uranus (only Cordelia resides within the ring system), so the smaller mass of the Neptune rings is not a consequence of less available source material. The reasons for the differences between the two ring systems are not understood, but they imply significant differences in the evolution of these two ring systems.

## 8.7 NOTES AND REFERENCES

[1] Porco, C. C., Nicholson, P. D., Cuzzi, J. N., Lissauer, J. L., Esposito, L. W., 1995, "Neptune's ring system", in *Neptune and Triton*, edited by D. P. Cruikshank, pp. 703–804.

[2] Miner, E. D., Wessen, R. R., 2002, *Neptune: The planet, rings and satellites*, chapters 3 ("Speculation about Neptune's rings", pp. 27–34) and 11 ("The rings of Neptune", pp. 237–246).

[3] Image reproduced from fig. 3 of Porco *et al.* [1], provided originally by M. Showalter.

[4] Tyler, G. L., Sweetnam, D. N., Anderson, J. D., Borutzki, S. E., Campbell, J. K., Eshleman, V. R., Gresh, D. L., Gurrola, E. M., Hinson, D. P., Kawashima, N., Kursinski, E. R., Levy, G. S., Lindal, G. F., Lyons, J. R., Marouf, E. A., Rosen, P. A., Simpson, R. A., Wood, G. E., 1989, "Voyager radio science observations at Neptune and Triton", *Science* **246**, 1466–1473.

[5] Lane, A. L., West, R. A., Hord, C. W., Nelson, R. M., Simmons, K. E., Pryor, W. R., Esposito, L. E., Horn, L. J., Wallis, B. D., Buratti, B. J., Brophy, T. G., Yanamandra-Fisher, P., Colwell, J. E., Bliss, D. A., Mayo, M. J., Smythe, W. D., 1989, "Photometry from Voyager 2: Initial results from the Neptunian atmosphere, satellites and rings", *Science* **246**, 1450–1454.

[6] Broadfoot, A. L., Atreya, S. K., Bertaux, J. L., Blamont, J. E., Dessler, A. J., Donahue, T. M., Forrester, W. T., Hall, D. T., Herbert, F., Holberg, J. B., Hunten, D. M., Krasnopolsky, V. A., Linick, S., Lunine, J. L., McConnell, J. C., Moos, H. W., Sandel, B. R., Schneider, N. M., Shemansky, D. E., Smith, G. R., Strobel, D. F., Yelle, R. V., 1989, "Ultraviolet spectrometer observations of Neptune and Triton", *Science* **246**, 1459–1465.

[7] Horn, L. J., Hui, J., Lane, A. L., Colwell, J. E., 1990, "Observations of neptunian rings by Voyager photopolarimeter experiment", *Geophysical Research Letters* **17**, 1745–1748.

[8] Broadfoot *et al.*, 1989, [6].

[9] Horn *et al.*, 1990, [7].

[10] Gurnett, D. A., Kurth, W. S., Granroth, L. J., Allendorf, S. C., Poynter, R. L., 1991, "Micron-sized particles detected near Neptune by the Voyager 2 Plasma Wave Instrument", *Journal of Geophysical Research* **96**, 19177–19186.

[11] Pedersen, B. M., Meyer-Vernet, N., Aubier, M. G., Zarka, P., 1991, "Dust distribution around Neptune: Grain impacts near the ring plane measured by the Voyager Planetary Radio Astronomy experiment", *Journal of Geophysical Research* **96**, 19187–19196.

[12] Stone, E. C., Miner, E. D., 1991, "The Voyager encounter with Neptune", *Journal of Geophysical Research* **96**, 18903–18906.

[13] Fig. 5 of Porco *et al.*, 1995, [1], found on p. 715 of that article.

[14] Smith, B. A., Soderblom, L. A., Banfield, D., Barnet, C., Basilevsky, A. T., Beebe, R. F., Bollinger, K., Boyce, J. M., Brahic, A., Briggs, G. A., Brown, R. H., Chyba, C., Collins, S. A., Colvin, T., Cook, A. F. H., Crisp, D., Croft, S. K., Cruikshank, D., Cuzzi, J. N.,

Danielson, G. E., Davies, M. E., De Jong, E., Dones, L., Godfrey, D., Goguen, J., Grenier, I., Haemmerle, V. R., Hammel, H., Hansen, C. J., Helfenstein, C. P., Howell, C., Hunt, G. E., Ingersoll, A. P., Johnson, T. V., Kargel, J., Kirk, R., Kuehn, D. I., Limaye, S., Masursky, H., McEwen, A., Morrison, D., Owen, T., Owen, W., Pollack, J. B., Porco, C. C., Rages, K., Showalter, M., Sicardy, B., Simonelli, D., Spencer, J., Sromovsky, B., Stoker, C., Strom, R. G., Suomi, V. E., Synott, S. P., Terrile, R. J., Thomas, P., Thompson, W. R., Verbiscer, A., Veverka, J., 1989, "Voyager 2 at Neptune: Imaging science results", *Science* **246**, 1422–1449.

[15] Showalter, M. R., Cuzzi, J. N., 1992, "Physical properties of Neptune's ring system", *Bulletin of the American Astronomical Society* **24**, 1029 (abstract); quoted in Porco *et al.*, 1995, [1], p. 715.

[16] De Pater, I., Gibbard, S. G., Chiang, E., Hammel, H. B., Macintosh, B., Marchis, F., Martin, S. C., Roe, H. G., Showalter, M., 2005, "The dynamic Neptunian ring arcs: Evidence for a gradual disappearance of Liberté and resonant jump of Courage", *Icarus* **174**, 263–272.

[17] Lindblad resonances are named for Swedish astronomer Bertil Lindblad (1895–1965). He was the first to recognize that galactic spiral structure could be attributed to the gravitational influence of a perturbing body on a disk of smaller bodies if the two orbited around a common center in periods that are related by relative small integers. For the purposes of this chapter, the three bodies involved in Lindblad resonances are the planet, a perturbing satellite of sufficient mass, and a ring. For example, if a satellite orbits a planet interior to a planetary ring and completes two orbits of the planet for every single orbit of some of the particles in the ring, and if the satellite is large enough to exert a measurable gravitational force from that distance, those particles in the ring will be perturbed by a 1:2 outer Lindblad resonance due to repeated gravitational tugs in the same direction. If the satellite is exterior to the ring and orbits the planet three times for every four orbits of the ring material, it might exert a perturbing force on the ring material at the 4:3 inner Lindblad resonance. In general, the smaller the satellite, the closer it must be to the ring and the higher the order of the Lindblad resonance. In the case of Galatea and the Adams ring, the latter is at a 42:43 outer Lindblad resonance radius of the former.

[18] Zarka, P., Pedersen, B. M., Lecacheux, A., Kaiser, M. L., Desch, M. D., Kurth, W. S., 1995, "Radio emissions from Neptune", in *Neptune and Triton*, edited by D. P. Cruikshank, pp. 341–387.

## 8.8 BIBLIOGRAPHY

Miner, E. D., Wessen, R. R., 2002, *Neptune: The planet, rings and satellites*, Springer-Praxis (note especially chapters 3 ("Speculation about Neptune's rings", pp. 27–33), 11 ("The rings of Neptune", pp. 237–246), and 14 ("Comparative planetology of the four giant planets", pp. 273–288).

Porco, C. C., Nicholson, P. D., Cuzzi, J. N., Lissauer, J. L., Esposito, L. W., 1995, "Neptune's ring system", in *Neptune and Triton*, edited by D. P. Cruikshank, pp. 703–804.

## 8.9 PICTURES AND DIAGRAMS

Figure 8.1 Fig. 3 of Porco *et al.* [1].
Figure 8.2 Fig. 5 of Porco *et al.* [1].

# Color section

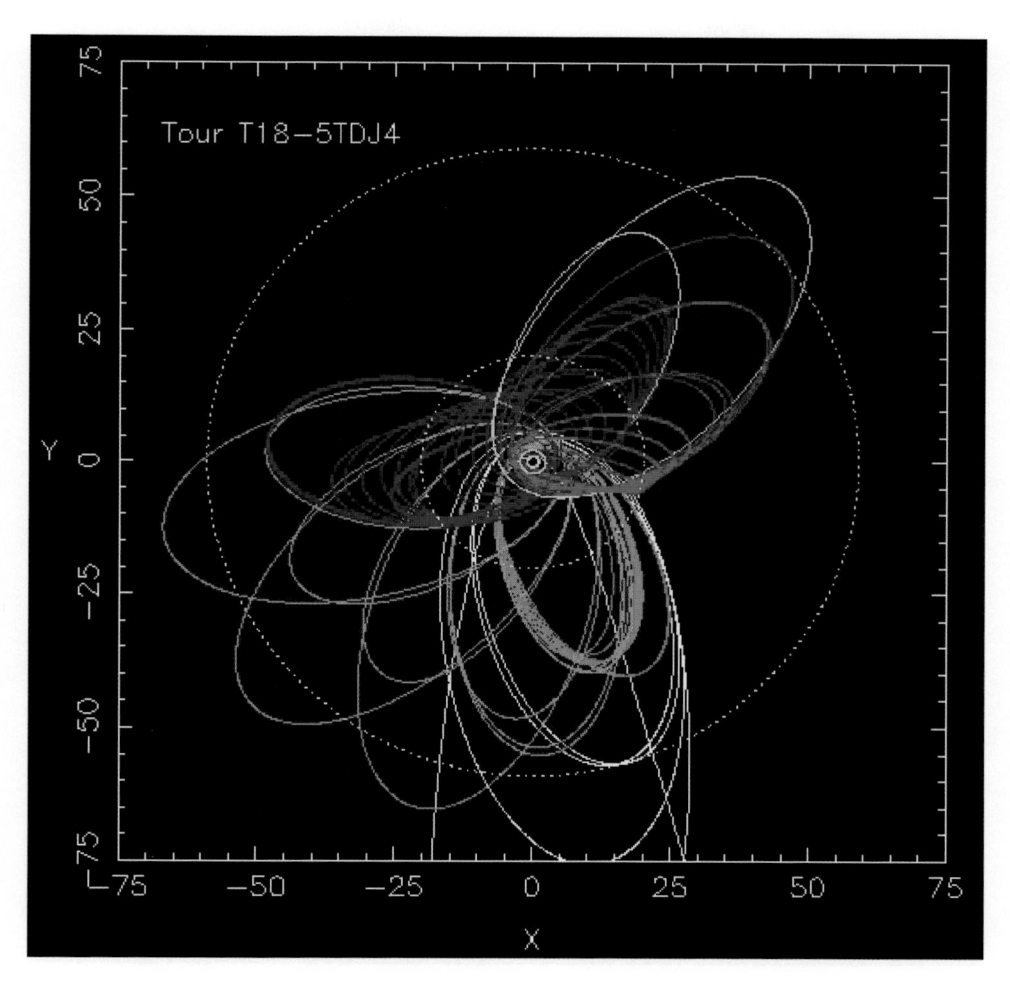


**Figure 10.3.** Cassini's 4-year tour is shown from Saturn's north pole. The colors of the orbits make it easier to identify which are which. During 2004, Cassini approached on the white trajectory and executed several white orbits in the fall and winter, dropping the Titan probe. After a few more equatorial orbits, it went into the orange series of inclined "ring orbits" during the spring and summer of 2005. The green equatorial orbits followed, until the fall of 2006, when the blue series of inclined orbits is planned to execute, followed closely by the red series which cranks up to very high inclination angles.

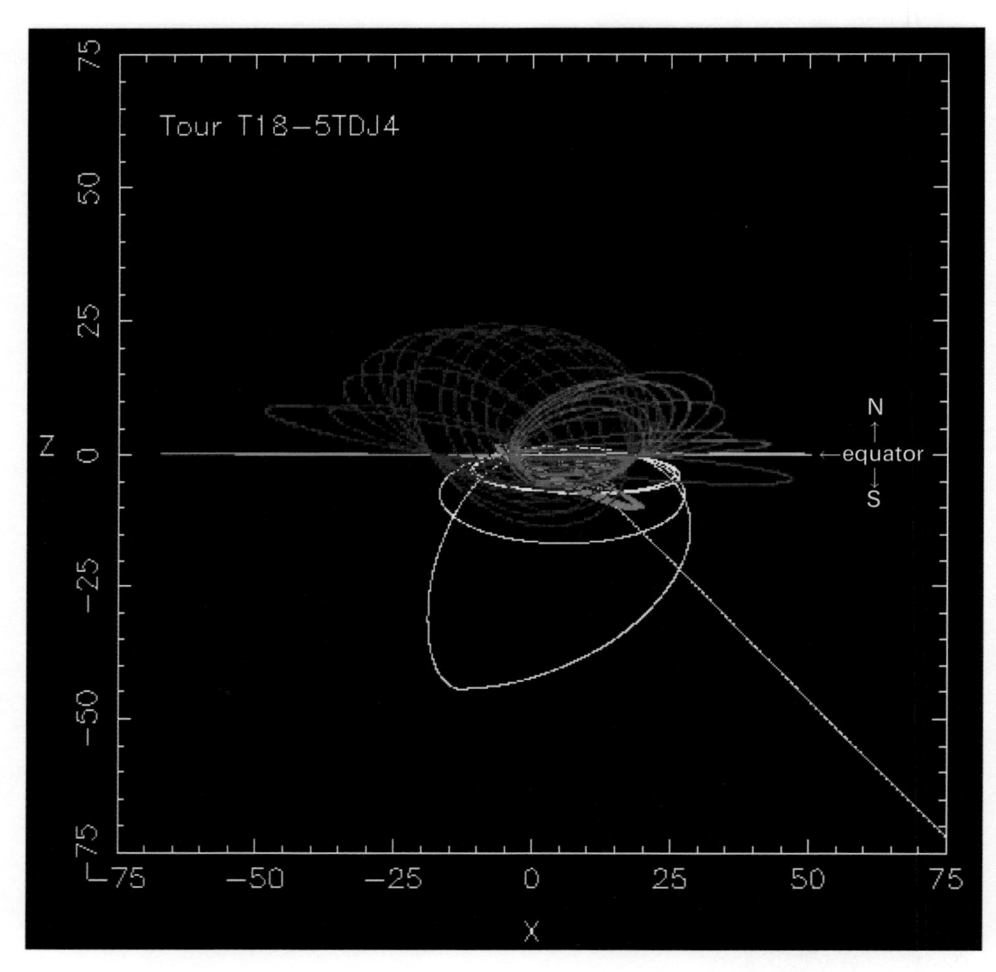


**Figure 10.3** (*cont.*). Cassini's 4-year tour is shown from Saturn's equator plane.

**Figure 10.18.** False color image of the planet and rings made by VIMS. Colors are selected to represent composition, so the rings, composed mainly of water ice, are blue and the planet, with its methane absorption is yellow (high clouds, less methane) or orange (deeper clouds, more methane. In the lower panel a red color is assigned to "organic" features, which here shows primarily light reflected from Saturn's bright noon hemisphere (onto the rings to the left) or transmitted through its upper atmosphere at the edges of its shadow (right side).

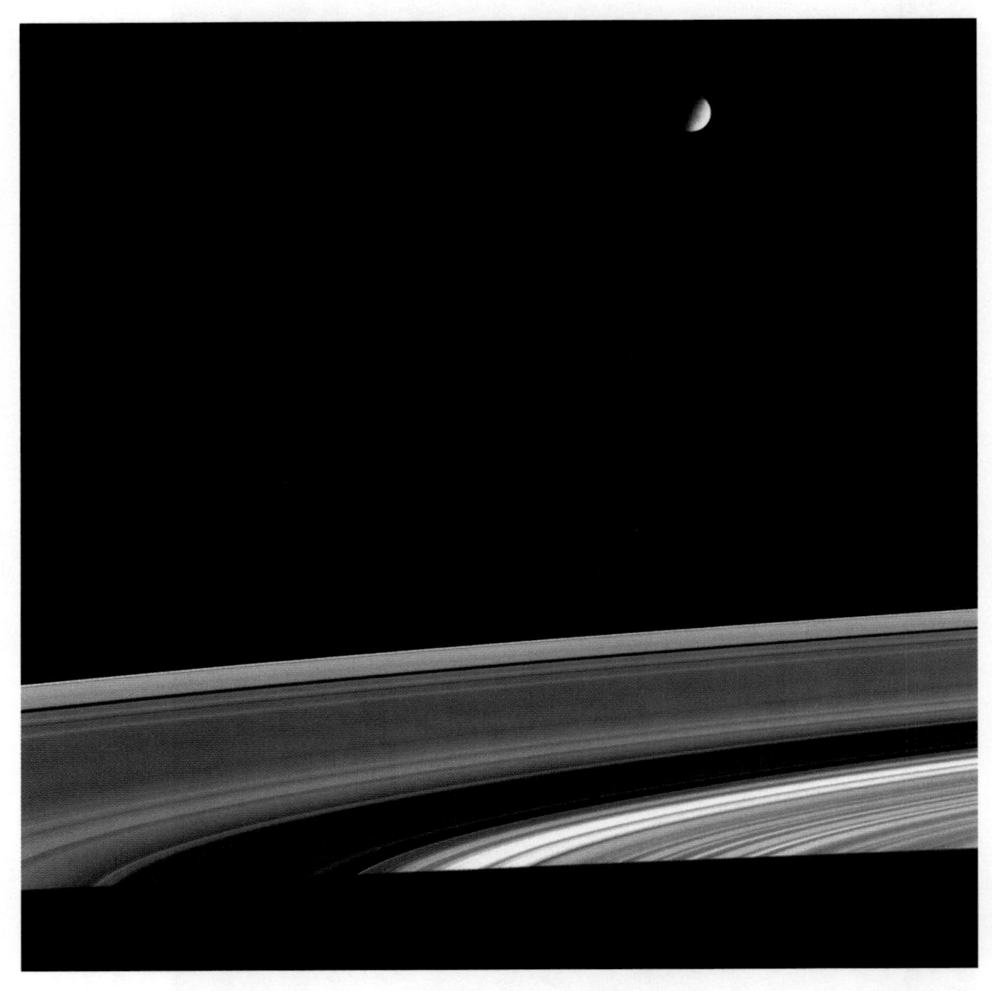

**Figure 10.19.** A direct comparison of the tawny colored rings and the nearly pure white, water ice surface of Enceladus. Some other material besides water ice is providing the reddish color in the rings, which is absent from Enceladus's surface. It has been suggested that some form of organic material, common in the outer solar system, might be providing the rings' coloration.

**Figure 10.20.** A color mosaic of the rings, intended to represent their true color. Subtle color variations can be detected in the B ring, with some ringlets looking less red than others.

**Figure 10.21 (top) and 10.22 (bottom).** Computer-generated "images" of the rings using RSS 3-wavelength occultation data, which really sample only one trace radially through the rings. Regions dominated by small particles are colored green and blue; regions dominated by large particles are colored red. Top: the full ring profile, with white regions being places which were so opaque that no size information has yet been extracted. Bottom: focus on the Cassini Division and A ring. Note the increasing abundance of small particles in the outermost A ring (blue color).

# 9

# Pre-Cassini knowledge of the Saturn ring system

## 9.1 INTRODUCTION

In Chapter 2 we summarized observations of Saturn's rings from the time of Galileo's first telescopic observations in 1610 through the end of the Voyager 2 encounter (1981). In this chapter we will discuss the scientific analysis of the Voyager (and subsequent Earth-based data) up to June 1, 2004, a month before the insertion of the Cassini Orbiter and the attached Huygens probe into orbit around Saturn on July 1. In the next chapter (Chapter 10), we will discuss the early findings of the Cassini Orbiter relative to Saturn's ring system. We purposely separate the two sets of observations, in part to emphasize the incredible rate with which understanding of the Saturn ring system (and indeed ring processes in general) is growing. As is often the case, Nature conspires against our complete understanding of the processes operative within the Saturn ring system. While additional observations improve our data set and answer many of our *a priori* questions, they also reveal myriads of details that call for new explanations. But then, such is the nature of scientific exploration. As unsettling as the thought might be to some readers of this book, we may never completely understand the processes operating within Saturn's rings, but we are certainly obtaining a more complete description of their physical characteristics and composition, and our understanding, while meager, continues to grow.

In Section 9.2 we summarize the radial dimensions, orbital speeds and orbital periods, and other radial characteristics of the rings. Vertical structure of the rings, including both physical and optical (including microwave) thicknesses is discussed in Section 9.3. Section 9.4 discusses the third dimension of the rings, namely their azimuthal structure. In Section 9.5 we discuss some of the gravitational mechanisms operating within the rings, along with the tangible outcomes of those interactions. Clear gaps within the rings and eccentric (non-circular) ringlets within those gaps are probably related to gravitational interactions, either with known or yet-to-be-discovered moonlets, but this interaction is not well understood; these are discussed in

Section 9.6. In Section 9.7 we discuss the gases that surround the rings and form an extremely tenuous ring atmosphere. The rings, of course, are actually collections of particles of varying size and composition; while there are no data that resolve the individual ring particles, much can be inferred about individual particle characteristics from the scattering properties of the rings, as we discuss in Section 9.8. Section 9.9 covers other optical properties of the rings, including color variations and variations in ring brightness with solar phase angle. Thermal properties of the rings are discussed in Section 9.10. Voyager radio data and its implications are discussed in Section 9.11. Evidence for interactions of the rings with Saturn's magnetic field is reviewed in Section 9.12. In Section 9.13 we discuss Earth-based observations of the rings since the Voyager encounters. We summarize in Section 9.14 some of the major remaining unanswered questions before consideration of Cassini data. Finally, in Section 9.15, we briefly outline some of the major scientific objectives for ring observations by the Cassini Orbiter.

## 9.2 NOMENCLATURE AND RADIAL DIMENSIONS OF THE SATURNIAN RINGS

In Chapter 2, the discovery of Saturn's rings was discussed in detail. A general qualitative description of the rings was included with that discussion, as was the nomenclature of the individual ring regions. In Table 9.1, adapted from Cuzzi *et al.* [1], the orbital characteristics and estimated masses of the rings are given. For comparison, we also include the inner eight satellites of Saturn known prior to Cassini.

The narrow F ring was first discovered by Pioneer 11 during its encounter with Saturn in 1979. The narrow rings of Uranus had been discovered two years earlier, so narrow rings were not entirely new, but their confining mechanisms were not yet well understood. So when Voyager 1 discovered two small satellites, Prometheus and Pandora, flanking the F ring, it was natural to assume their presence was somehow responsible for gravitationally confining the ring material between them to a very narrow radial range. They became the archetypical shepherding satellites (Figure 9.1). It was not long before ring scientists recognized that the gravitational interaction of these two shepherds with the F ring was not nearly as orderly as at first supposed, for the F ring had many irregular structures, including kinks, brighter clumps, and what appears to be braiding (Figure 9.2).

Aside from the irregularities at the outer edge of the B ring and in the F ring, the Saturn ring system seems to have a relatively orderly radial structure. Three types of observations confirm that general conclusion. In addition to the images from Voyagers 1 and 2, Voyager 1 conducted a two-frequency radio occultation scan of the rings, and Voyager 2 conducted a two-instrument stellar occultation scan of the rings. In its radio occultation experiment, Voyager 1 radiated S-band (wavelength 13 cm) and X-band (3.6 cm) radio signals toward Earth as the spacecraft passed behind Saturn's rings as viewed from Earth. The signals were received by large radio dish antennas on Earth. In each of the stellar occultation measurements, Voyager 2's photopolarimeter (a sensitive light meter) and ultraviolet spectrometer observed the

**Table 9.1.** Saturn ring dimensions ($1R_S = 60{,}330$ km, $1M_S = 5.685 \times 10^{26}$ kg).

| Ring region | Boundary (km) | Boundary ($R_S$) | Velocity (km/s) | Period (hr) | Mass ($M_S$) |
|---|---|---|---|---|---|
| D ring | 66,900 | 1.11 | 23.78 | 4.91 | ?? |
| | 74,510 | 1.235 | 22.56 | 5.76 | |
| C ring | 74,510 | 1.235 | 22.56 | 5.76 | $\sim 2 \times 10^{-9}$ |
| | 92,000 | 1.525 | 20.30 | 7.91 | |
| B ring | 92,000 | 1.525 | 20.30 | 7.91 | $\sim 5 \times 10^{-8}$ |
| | 117,580* | 1.949* | 17.97 | 11.42 | |
| Cassini Division | 117,580* | 1.949* | 17.97 | 11.42 | $\sim 1 \times 10^{-9}$ |
| | 122,170 | 2.025 | 17.62 | 12.10 | |
| A ring | 122,170 | 2.025 | 17.62 | 12.10 | $\sim 1.1 \times 10^{-8}$ |
| *(Pan* | *133,580* | *2.214* | *16.89* | *13.80* | $\sim 1.1 \times 10^{-10}$*)* |
| | 136,780 | 2.267 | 16.66 | 14.33 | |
| *(Atlas* | *137,640* | *2.282* | *16.62* | *14.45* | $\sim 1.5 \times 10^{-11}$*)* |
| *(Prometheus* | *139,350* | *2.310* | *16.52* | *14.72* | $\sim 2.3 \times 10^{-10}$*)* |
| F ring | 140,180* | 2.324* | 16.49 | 14.84 | ?? |
| *(Pandora* | *141,700* | *2.349* | *16.38* | *15.10* | $\sim 2.6 \times 10^{-10}$*)* |
| *(Epimetheus* | *151,422* | *2.510* | *15.87* | *16.65* | $\sim 8 \times 10^{-10}$*)* |
| *(Janus* | *151,472* | *2.511* | *15.84* | *16.69* | $\sim 2 \times 10^{-9}$*)* |
| G ring | 162,000 | 2.69 | 15.32 | 18.45 | ?? |
| | 175,000 | 2.90 | 14.74 | 20.73 | |
| E ring | 181,000 | 3.0 | 14.5 | 21.8 | ?? |
| *(Mimas* | *185,520* | *3.075* | *14.32* | *22.61* | $6.6 \times 10^{-8}$*)* |
| *(Enceladus* | *238,020* | *3.945* | *12.64* | *32.87* | $1.5 \times 10^{-7}$*)* |
| | 483,000** | 8.0** | 9.7 | 87.3 | |

* The outer edge of the B ring is non-circular; the F ring is azimuthally non-uniform and time-variable. Estimated mean values are given in the table.

** Cassini data seem to indicate that the E ring extends much further out, perhaps nearly to Titan, which orbits at a radial distance of 1,221,830 km = 20.252 $R_S$ (see Figure 2.15 and Chapter 10).

light of a star, which, due to the motion of Voyager 2, appeared to pass behind the rings. These occultation profiles are best displayed by calculating from the measurements the amount of ring material needed either to block a portion of the radio signal received at Earth or to reduce the apparent brightness of the starlight measured by the two Voyager optical instruments.

Here it is useful to re-introduce the concept of optical depth. An object (or portion of a ring) that completely blocks the radio signal or starlight is defined to have an infinite optical depth; if there is no reduction in the radio signal strength or the apparent star brightness, the object has zero optical depth. For most of the ring occultation measurements, the measured optical depth is greater than 0 and less than infinite. Expressed as a mathematical equation, the intensity of the measured radio

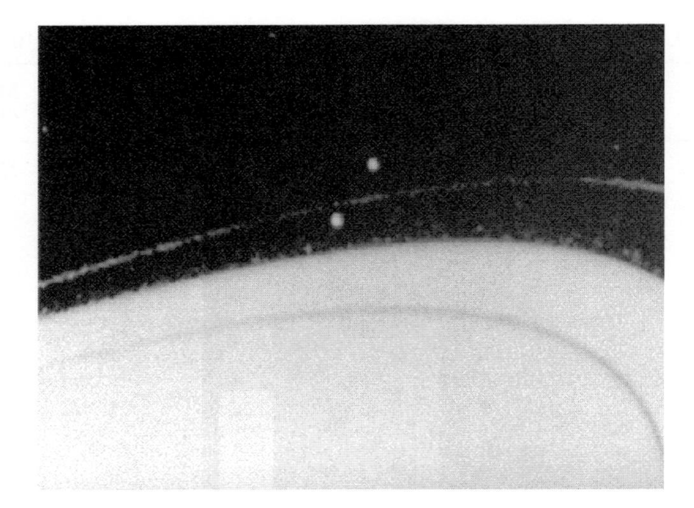

**Figure 9.1.** Voyager 1 captured this image in 1980 which shows the F ring and its two shepherding satellites, Prometheus (inner) and Pandora (outer). (P-23911)

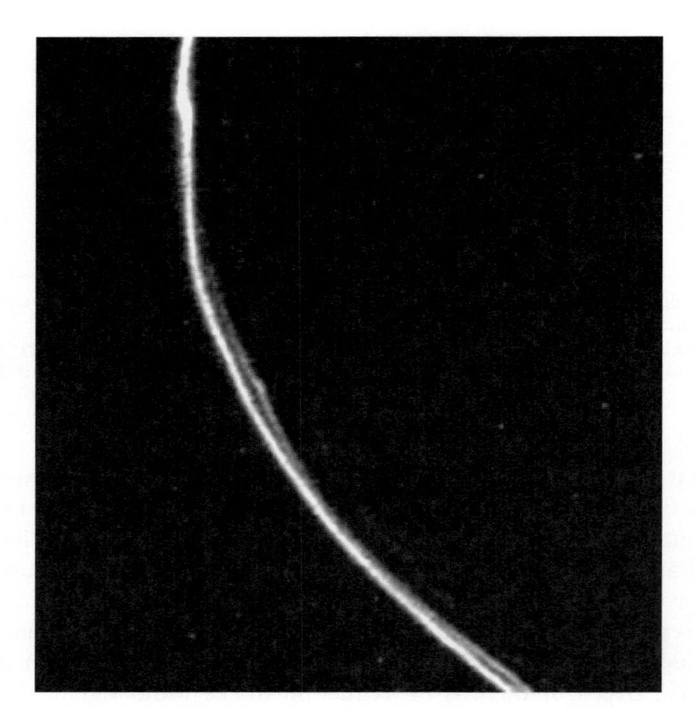

**Figure 9.2.** This view of the F ring was shuttered by Voyager 1 from a range of 750,000 km and shows kinks, clumps, and apparent braiding within the F ring, apparently as a result of complex gravitational interactions with its two shepherding satellites. (PIA02283)

beam or starlight, **I**, is related to the normal unblocked intensity, $\mathbf{I_0}$, by the following relationship:

$$\mathbf{I} = \mathbf{I_0}\,\mathbf{e}^{-\tau},$$

where $\tau$ is the optical depth of the intervening material and $\mathbf{e} \approx 2.718\,28$ (**e** is the so-called base of natural logarithms). If the optical depth $\tau = 0$, then $\mathbf{I} = \mathbf{I_0}$, and there is no reduction in the intensity of the signal. If $\tau = \infty$, then $\mathbf{I} = 0$, and the signal is entirely blocked. If the optical depth $\tau = 1$, the signal intensity is reduced by a factor of $1/\mathbf{e}$, or to a level approximately 36.8% of the unblocked intensity. For an optical depth $\tau = 2$, the signal intensity is reduced by a factor of $1/\mathbf{e}^2$, or to approximately 13.5% of the unblocked intensity, and so forth.

One additional factor that must be taken into account is the angle the ray (from the star to Voyager 2 or from Voyager 2 to Earth) makes with the plane of the rings. A vertical ray path through the rings will encounter the fewest possible ring particles during its passage. An oblique ray that is tilted from the ring plane by only a few degrees will generally encounter more particles during its passage through the ring. Optical depth determinations are often corrected for this factor. The simplest correction is to multiply the measured optical depth by the cosine of the angle between the ray and the vertical to the ring plane. The cosine of 0° is 1; the cosine of 60° is $\frac{1}{2}$. For the radio occultation experiment on Voyager 1, the radio beam was only 5° above the ring plane, which corresponds to 85° from vertical; the cosine of 85° is only 0.087.

There are, however, two circumstances for which a simple cosine correction will not provide a good estimate of vertical optical depth. If the ring is essentially a single particle thick, the variation of apparent optical depth with the angle from the vertical has a different form, allowing us to infer vertical structure from measurements at different opening angles. In such a situation, the projected area of an isolated ring particle does not change, but the projected area of the empty space between the particles is very dependent on the spatial density of the particles and on the ring tilt angle, especially for small tilt angles [2]. On the other hand, if the ring is not perfectly flat, but is, for example, corrugated due to the gravitational influence of a satellite in a slightly inclined orbit, it may not be possible to express the measured optical depths as vertical optical depths. Both of these circumstances may exist at places in the Saturn ring system. Corrugations (bending waves) have been observed in the A ring, and there is reason to believe that the C, B, Cassini Division, and A rings may not be markedly thicker than the diameters of the larger particles making up those rings (see Section 9.3).

Typical measured optical depths (uncorrected for non-vertical viewing) in the C ring are about 0.1, with some of the narrow ringlets reaching optical depths of about 0.4. Ring gaps, including the Maxwell gap near a radius of 87,500 km in the C ring, the Huygens gap at the outer edge of the B ring (radius of about 117,680 km), the Encke gap centered near 133,570 km in the A ring, and the Keeler gap near 136,530 km radius in the A ring, have essentially 0 optical depth over most of their widths, which range from a few tens of kilometers to several hundred kilometers. Except for the inner third of the B ring, the entire B ring has an optical depth in excess of 1.0, and could not be easily distinguished from infinite optical depth in many of the measurements. The

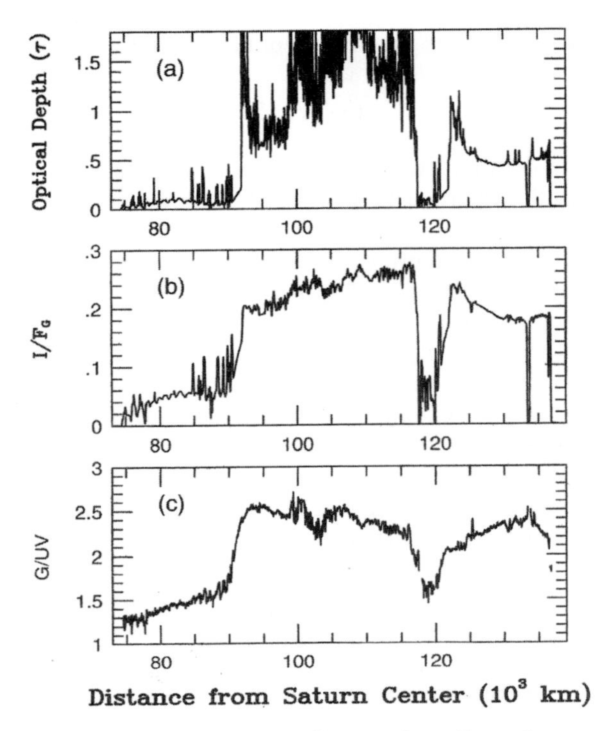


**Figure 9.3.** Ring characteristics versus radial distance from Saturn's center (in thousands of km). (a) Optical depth as determined from stellar occultation data, showing the main ring regions. (b) Ring reflectivity in green light as determined from Voyager imaging data. (c) Ratio of reflectivity in Voyager imaging green light to that in the ultraviolet light. (Figure from Estrada *et al.* [3])

mean optical depth of the Cassini Division was generally less than 0.1, except for a couple of ringlets of optical depth near 0.5 in the outer half and a ramp at the outer edge that reached a maximum optical depth near 0.2. The A ring optical depth was generally near 0.5 except near its inner edge, where the optical depth exceeded 1.0. Figure 9.3(a) displays in graph form the measured optical depths versus radial distance from Saturn's center for one of the stellar occultations as determined by Voyager's photopolarimeter. The figure is from Estrada *et al.* [3]. Figure 9.3(b) shows the intensity of sunlight (in green light) reflected from the illuminated face of the rings, as determined by Voyager's imaging system. Note the similarity between boundaries and ringlet locations in Figures 9.3(a) and 9.3(b). Figure 9.3(c) shows the relative color of the rings as a plot of green brightness divided by ultraviolet brightness, again from Voyager imaging. Note that the C ring and the Cassini Division are considerably less red than the A and B rings, implying small differences in composition between these rings.

Optical depths of the other rings (D, G, and E) are not as well determined, since none were detected by the occultation measurements, and only portions of the F ring were detected in stellar and radio occultations. The F-ring optical depth is about 0.1 at its densest part. The G-ring optical depth is about 0.000 001; the D-ring optical depth is comparable with that of the G ring. The E-ring optical depth varies with radial distance from about 0.000 01 near its inner edge to much smaller values at greater distances.

## 9.3 VERTICAL STRUCTURE IN SATURN'S RINGS

The rings lie in Saturn's equatorial plane. That equatorial plane is tilted by 26.7° to the plane of Saturn's orbit around the Sun. The tilt results in seasons on the ringed planet similar to those on Earth. On Earth, the apparent path of the Sun crosses our equator every six months, on or close to March 20 and September 20. These crossings are known, respectively, as the vernal and autumnal equinoxes. Saturn circles the Sun in 29.4 years; therefore, every 14.7 years, it also has equinoxes, during which times its rings are edge-on to the Sun. The orbital motion of the Earth is much faster than the orbital motion of Saturn, and the Earth's orbital plane is not exactly the same as Saturn's orbital plane; one consequence of these facts is that viewers on Earth can sometimes view Saturn's rings edge-on three times in relatively quick succession near the times of Saturn's equinox. A triple crossing occurred in the 1995–6 time period, as shown in Figure 9.4. As viewed from Earth, the 2009 Saturn ring plane crossings will be only a single event; the next triple crossing will occur in 2023.These edge-on presentations offer Earth-bound observers the opportunity to study a variety of phenomena at Saturn, including studies of small satellites swamped by the brightness of the rings at other times, studies of faint rings that are otherwise too faint to be seen, and studies of the vertical thicknesses of all the rings of Saturn.

The thickness of the main rings cannot be directly observed from Earth, but if one plots their photometric brightness at a variety of tilt angles and extrapolates that brightness to zero tilt angle, the main rings have an effective photometric thickness of about 1 km [4]. This is probably an overestimate of their thickness, since there are several factors that might tend to make the effective photometric thickness much larger than their physical thickness. First, the photometric brightness includes contributions from the F, G, and E rings in front of and behind the main rings [5]. Also, as mentioned earlier, the corrugations (bending waves) in parts of the A ring due to the gravitational influence of Mimas in its inclined orbit make the apparent thickness larger than the actual thickness of the rings. These corrugations have peak-to-peak amplitude on the order of 1 km [6]. Another contributor to the apparent thickness of the rings is slight warping of the ring plane due to long-term gravitational effects of the Sun and Saturn's largest satellite, Titan [7].

In addition to the observations from Earth, there are some measurements from Voyager 2 that bear directly on estimates of the thickness of the rings. The Voyager 2 photopolarimeter measurement of a stellar occultation had a radial resolution of about 100 meters in the rings. The outer edge of the A ring cut off the starlight nearly instantaneously, and from that fact the photopolarimeter team were able to place an upper limit of 200 m on the thickness of the outer edge of the A ring [8]. The radio

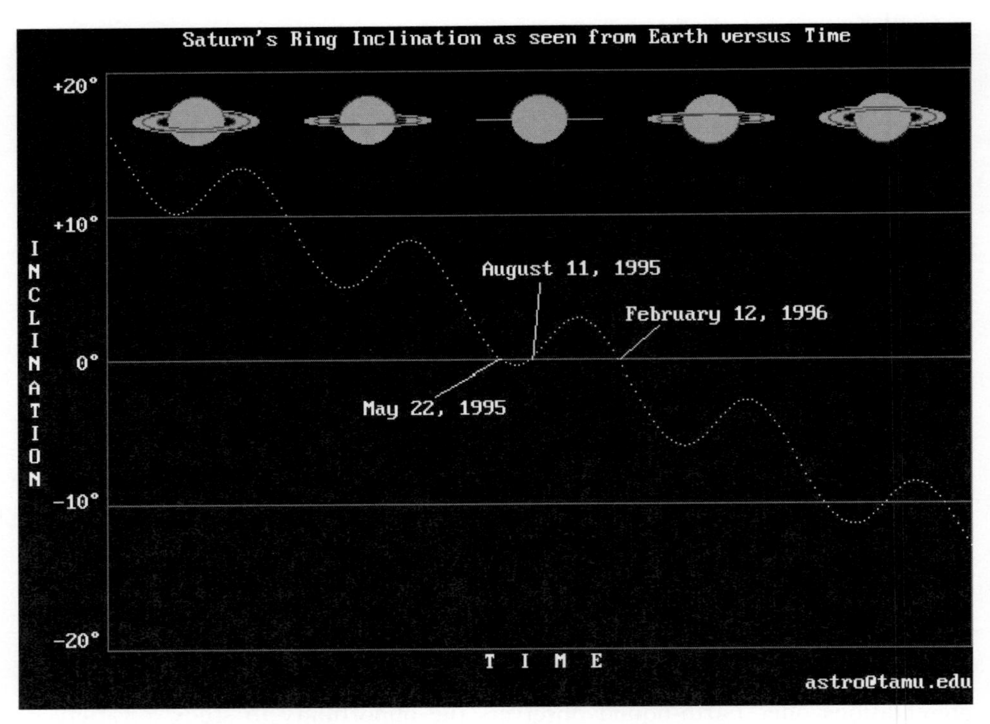


**Figure 9.4.** This diagram shows the Saturn ring opening angle versus time for the 1995–6 time period. The Earth crossed Saturn's ring plane on May 22 and August 11, 1995, and then again on February 12, 1996. The Sun crossed the ring plane on November 19, 1995. (Courtesy NASA/JPL)

occultation measurements of Voyager 1 utilized a coherent radio signal [9] at both the S-band and X-band. Because the signal was coherent, a sharp edge in the rings would create a diffraction pattern [10] that would sweep across the receiving station on Earth. The spacing of peaks in the diffraction pattern could then be used to estimate the thickness of the rings at each point where they occurred. The radio science team used these measurements to come up with upper limits of 150 to 200 m on the thickness of the main rings at several sharp ring edges [11]. Furthermore, the additional (gravitational?) energy needed to maintain a sharp edge would cause the edge of a ring to be thicker than nearby interior areas [12]. Combining these results with some theoretical considerations associated with the characteristics of wave structures within the rings (see Section 9.5) lead to the conclusion that the thickness of the main rings, at least away from their edges, is between 10 and 100 m.

The extremely low value for the ring thickness has renewed speculation about whether the main rings are essentially a monolayer—that is, they may only be a single particle in thickness. In such a situation, the particles, although consisting of a range of sizes, orbit the planet in very nearly a single plane at or near Saturn's equatorial plane. (More will be said about particle sizes and other particle characteristics in

Section 9.8.) Monolayer structure may even be preserved in the localized corrugated regions where bending waves are found and in the large-scale warping of the ring plane, although in such circumstances the particles would leave the precise equatorial plane.

There are several considerations which argue against a purely monolayer structure for the main rings. First, even if the ring particles initially were all in circular, non-inclined orbits, their physical sizes, combined with the fact that particles closer to the planet move with higher velocity than those further from the planet, would lead to occasional interparticle collisions. These collisions will result in non-circular, inclined orbits for the involved particles, increasing the likelihood of future collisions, and a certain amount of randomization of velocities will be maintained by this process, formally known as Keplerian shear. Because of their smaller masses, smaller particles will have their velocities dispersed by larger amounts than larger particles. Thus, because of this Keplerian shear, the rings will tend toward a thickness that is several, but not many, particles in thickness [13]. It is possible that by this process the smaller particles will form a somewhat greater ring thickness than the larger particles [14]. Another consequence of Keplerian shear is that most collisions will also impart a spin to the involved particles that is preferentially in a direction opposite that of their orbital motion, but of about the same angular rate.

Another observational fact that must be considered in this discussion is the opposition effect long seen in the rings, in which the rings become distinctly and non-linearly brighter as the phase angle approaches 0°. Phase angle is defined as the angular separation between the Sun and the observer as seen from the target. Earth's full Moon is so bright because of its strong opposition effect near zero phase angle, although a strictly zero phase occurs for Earth-bound observers only when the Moon is in Earth's shadow. This opposition brightening of Saturn's rings has been interpreted as a reduction in shadowing, either at a large scale (particles cover their own shadows) or at a small scale (particle surface roughness) [15]. The latter is not as likely, especially for particles as reflective as the icy particles in Saturn's ring. Mutual shadowing of ring particles requires that the rings be at least a few particles in thickness. However, other possibilities have recently been found more likely (see Chapter 10).

The radio occultation results imply a thin, but not quite monolayer, structure for Saturn's rings [16]. Radar signals transmitted from Earth and reflected by the rings of Saturn also display a tilt angle dependence that is inconsistent with a monolayer [17]. The wider open the rings are, the better they reflect the radar signal back to Earth, and the difference is larger than the projected area of the rings as seen from Earth.

On the basis of the above considerations, it appears that the main rings are extremely thin, less than 150 to 200 m at measured sharp edges and more likely 10 to 100 m in thickness away from those edges. Nevertheless, the thickness is at least several times as large as the diameters of the most numerous ring particles.

The thicknesses of the tenuous G and E rings, in contrast, are certainly much larger than 10 to 200 m. Earth-based observations during ring-plane crossings clearly establish the fact that the E ring has a vertical thickness of several thousand kilometers and may in fact increase in thickness with distance from Saturn [18]. These measurements also seem to show a local depression in thickness and an increase in the brightness near the orbit of the satellite Enceladus. That, combined with evidence from the Voyager images of a geologically young surface for Enceladus, has fueled speculation that Enceladus may be the source of, or at least a primary contributor to, the material in the E ring (see Chapter 10).

Pioneer 11 and Voyagers 1 and 2 passed through portions of the E and G rings. Each had instrumentation that was able to sense impacts on the spacecraft of large numbers of the micron-sized particles that make up these rings. From the time history of several of these ring passages, investigators have been able to estimate the effective thickness of about 2,000 km for the E ring [19], in reasonable agreement with ground-based measurements, and of about 100 km near the outer edge of the G ring (at 2.88 $R_S$) [20].

The physical thicknesses of the D and F rings are not well determined. The orbits of Prometheus and Pandora, which flank the F ring, are both slightly elliptical, and their interaction with the F ring leads to an irregular, non-circular shape. It is possible that both the D and F rings are of the same order of thickness as the A, B, and C rings, about 10 to 100 m, or they could be somewhat larger.

## 9.4 AZIMUTHAL NON-UNIFORMITY IN SATURN'S RINGS

The particles in the main rings pass through Saturn's shadow every orbit of the planet, except when the rings are wide open (near the times of summer and winter solstices). Near Saturn's solstices, particles in the A ring are continuously illuminated (see Figure 9.5). The shadow of Saturn on the rings is difficult to view from Earth-based telescopes because is it generally hidden by the planet. Figure 9.6 shows a view from Voyager 1, when the rings had a very low tilt angle; note that the shadow of Saturn consequently hides a relatively equal portion of each of the main rings. Passage of ring particles through Saturn's shadow causes a drop in temperature of a few degrees. Temperature is one of the few characteristics of the rings observable from Earth that varies with ring longitude. Other thermal characteristics of Saturn's rings will be covered in Section 9.10.

One other subtle longitudinally variable characteristic of the rings is easy to describe but much more difficult to explain. The two sides of the ring that extend outward (left and right or east and west) from Saturn are referred to as the ring ansae, a Latin word meaning cup handles. Each ansa (handle) has a near and a far part. Photometric brightness measurements show that the near portion of the left (east) ansa of the A ring is slightly brighter than its far portion. Similarly, the far part of the right (west) ansa of the A ring is slightly brighter than its near portion. This asymmetry was first reported by Camichel [21] and later confirmed by others [22]. Whatever is causing this asymmetry is apparently occurring at much smaller scales than are resolvable from ground-based or Voyager data. It seems to peak about 20°–25° before ring particles cross the observer-to-Saturn line, both on the near and far portions of the rings; the effect seems strongest near a ring tilt angle of about 12° [23]. Careful analysis of Voyager imaging [24] shows that the contrast between maximum

**Figure 9.5.** Saturn as seen from Cassini on November 9, 2003, when the rings were close to their widest opening. Cassini's range from Saturn was 111.4 million km. Note that the shadow of Saturn extends across all but the A ring. A view from this perspective cannot be captured by Earth-based telescopes due to the fact that the Earth is always within about 6° of the Sun as seen from Saturn. (PIA04913)

**Figure 9.6.** Saturn as seen from Voyager 1 on November 16, 1980, four days after closest approach. Voyager 1's range from Saturn is 5.3 million km. Compare the shape of Saturn's shadow across the rings with that of Figure 9.5. Here the tilt angle of the rings as viewed from the Sun is only about 5°. (PIA00335)

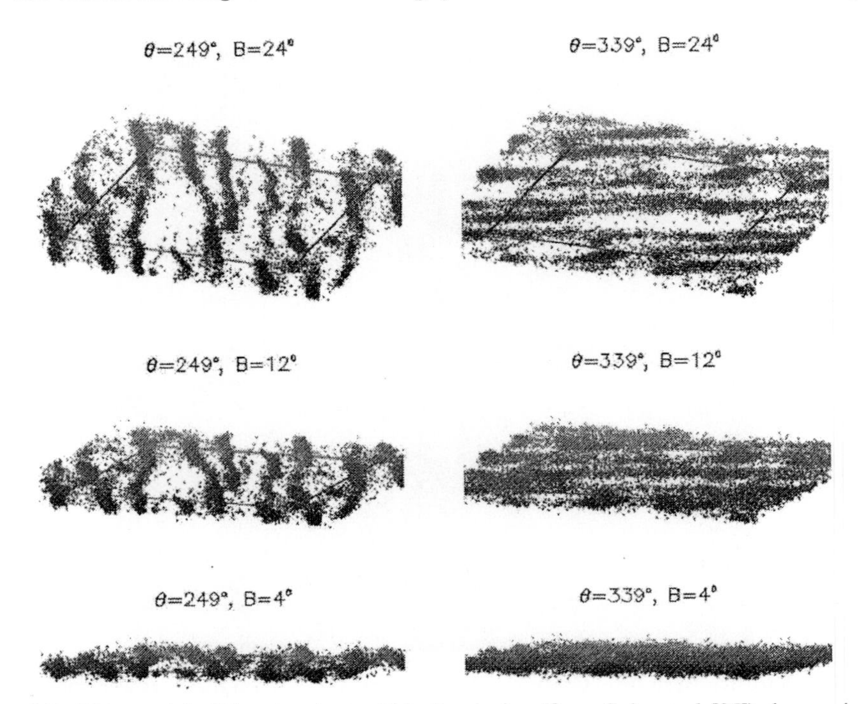


**Figure 9.7.** This model of the structure within the A ring (from Salo *et al.* [25]) shows views at 249° longitude (= 111° before meridian crossing) and at 339° longitude (= 21° before meridian crossing) at three different tilt angles. The sketch shows how the contrast can be greatest at the intermediate tilt angle. The filamentary appearance is thought to be loose collections of particles ("dynamical wakes") rather than individual elongated particles.

and minimum brightness amounts to about 35% of the mean brightness and is strongest near the center of the A ring. The Voyager 1 radio occultation experiment also noted some strange behavior in the data from the A ring [25]; the data seemed to indicate the presence of very elongated, aligned structures within the ring. Salo *et al.* [26], using data from the Hubble Space Telescope, published a possible theoretical explanation of the phenomenon, which involves temporary formation of approximately linear, tilted structures due to gravitational instabilities. Their results seem to be in reasonable agreement both with the positions of minima and maxima and with the dependence on tilt angle. The linear structures are broadside to the observer at 21° before passage of the ring particles across the Saturn–observer line and end-on 90° of ring longitude on either side of that peak (i.e., at 111° before "meridian" crossing and 69° after), as illustrated in Figure 9.7.

With the high resolution available from Voyager data, many more instances of azimuthal variability were discovered. The F-ring irregularities were mentioned briefly in Section 9.2 and depicted in Figure 9.2. We now know that F-ring structure changes on a very short timescale. There were changes in the appearance and location of clumps within the F ring in the time period between the Voyager 1 encounter

(November 12, 1980) and the Voyager 2 encounter (August 26, 1981) [27]. Some of the structure cannot be explained by the gravitational effects from the shepherding satellites, Prometheus and Pandora, which themselves undergo sporadic and rapid changes in their orbit parameters [28]. Additional data from the Cassini Orbiter were sorely needed to begin to understand this strange ring (see Chapter 10).

One of the major surprises of Voyager's ring studies was the discovery of enormous, nearly radial structures (spokes) in the outer half of the B ring. Viewed from low phase angles (with the Sun at the observer's back), features such as those in Figure 9.8 appear as dark markings on an otherwise bright B ring. When seen at high

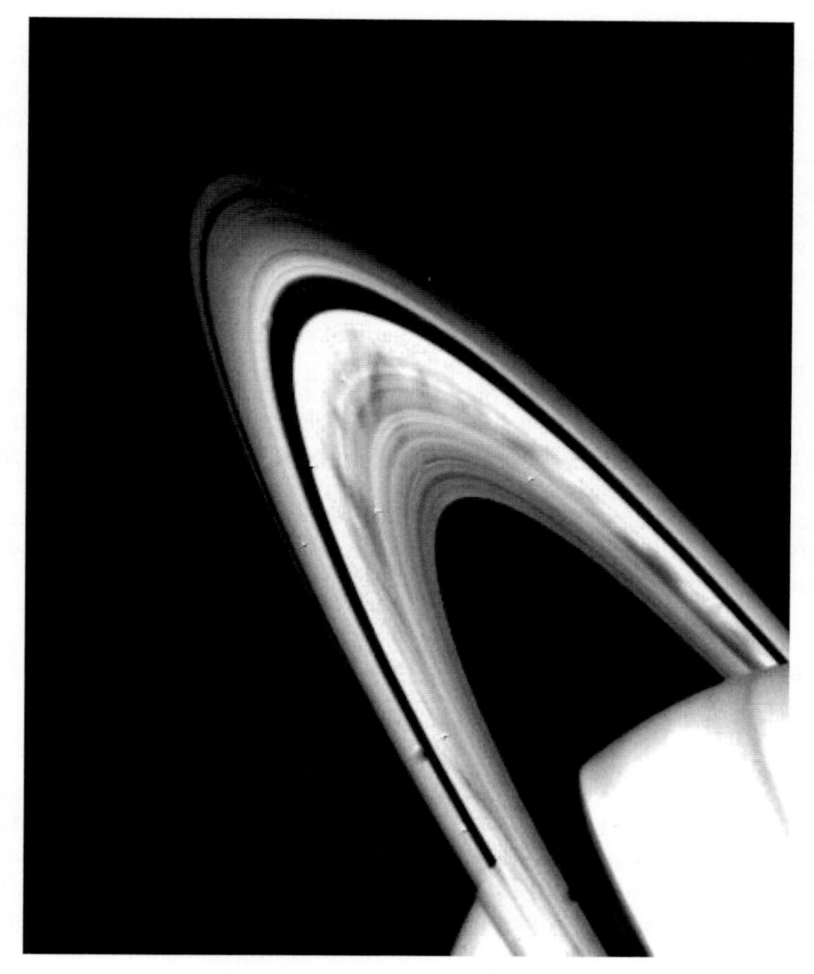

**Figure 9.8.** Dark, almost radial spokes are visible in the outer half of the broad B ring in this Voyager 2 image taken on August 3, 1981 (about three weeks before encounter) from a range of about 22,000,000 km. They are composed of tiny particles and are probably levitated above the main rings by electrostatic forces. (PIA02274)

phase angle (from the anti-Sun direction) against an illuminated B ring, the spokes appear brighter than the B ring. They also appear bright against the dark unilluminated side of the B ring. But the most startling feature of the spokes is that they seem almost entirely unaffected by Keplerian shear; that is, they maintain their shape in spite of the differential rotation occurring in the underlying B ring. In fact, they rotate, at least initially, at the rate of rotation of Saturn's magnetic field. They seem to appear almost instantaneously, generally (but not exclusively) in or near passage through Saturn's shadow, and last from less than one to as many as three times around the planet. By inspection of Table 9.1, it is easy to see that the period of rotation of Saturn ($\sim$10.656 hr) is approximately equivalent to the orbital period of B-ring particles about three-quarters of the way out in the ring. The reflective properties of the spokes are characteristic of tiny micrometer-sized particles. Perhaps these tiny particles are created by high-velocity meteoroid impacts with ring particles and soon become electrically charged (through the action of sunlight or by impacts with charged particles trapped in Saturn's magnetic field?). If these particles have low enough mass and high enough charge, they could become (at least temporarily) "frozen" in the rotating magnetic field, particularly if that differs only slightly from their normal orbital speed. In addition, electrostatic forces might tend to elevate these tiny particles above the main ring and therefore make them more visible to Voyager's cameras. It is safe to say that we still have much to learn about the nature and origin of and the processes acting within and upon the B-ring spokes. Recent observations of B-ring spokes using the Hubble Space Telescope [29] and observations by the Cassini spacecraft in orbit around Saturn (Chapter 10) are the first steps in that direction.

Other azimuthally variable features in the rings are in general due to gravitational effects. These include *wakes* at the inner and outer boundaries of gaps or rings, *density waves*, *bending waves*, and effects due to the *self-gravity* of the rings. These will be discussed in more detail in the following section, along with a number of other effects that may or may not be associated with gravitational effects, but cannot presently be associated with other known sources.

## 9.5 GRAVITATIONAL INTERACTIONS OF SATURN'S RINGS AND SATELLITES

Saturn has more "regular" satellites than any other planet in the solar system. Although some of these regular satellites (i.e., satellites with prograde, low-inclination, low-eccentricity orbits) may not be high-density, well-consolidated objects—most of them interact gravitationally with Saturn's ring system. The regular satellites include the classical satellites Mimas, Enceladus, Tethys, Dione, Rhea, Titan, Hyperion, and Iapetus. Prior to any new discoveries by Cassini, the set of regular satellites also included Pan, Atlas, Prometheus, Pandora, Janus, Epimetheus, Telesto, Calypso, and Helene. The last three of these are co-orbital with Tethys and Dione, whose effects completely mask those of their small companions. Hyperion is also small enough and distant enough that no gravitational effects of Hyperion on the rings have been noted. That still leaves at least 13 satellites that have each made some contribution to the structure of Saturn's rings. Apparently, Jupiter's ring structure is affected by only four satellites: Metis, Adrastea, Amalthea, and Thebe. Uranus has a total of 15 regular satellites, but gravitational interactions with the rings of Uranus have only been identified for a handful of these. Only 6 of Neptune's satellites are regular, and their interaction with the rings of Neptune is, if anything, even more poorly understood than that of the Uranus system. Thus, if we are to understand ring–satellite interactions, it is the Saturn ring system that will bring about that understanding. It is not only the scientifically richest natural laboratory for ring studies, it is the sole available natural laboratory that will contribute significantly to unraveling the enormously complex field of ring dynamics and evolution.

With the exception of some of the smaller satellites which undergo some relatively sudden changes to their orbits [30], all of the regular satellites of Saturn have fixed orbit periods (the time required for one circuit of Saturn). Even those which undergo changes have stable orbit periods between their brief chaotic episodes. Ring particles have orbit periods which vary with their distance from the planet's center, as can be seen by inspection of Table 9.1. When ring particles are at a distance where their orbit period is a simple fraction of that of a regular satellite (i.e., $\frac{1}{2}$, $\frac{1}{3}$, $\frac{2}{3}$, $\frac{1}{4}$, $\frac{3}{4}$, etc.), the ring particles are repeatedly nudged in the same direction and at the same point in their orbit. These fractional matches in orbit period are called *gravitational resonances*. Over time, these resonances result in slight alterations of the ring particle orbits. Initially, the tendency is to change the orbits from circular to elliptical; however, that also increases the frequency of collisions with nearby particles which resists the departure from circular orbits.

Several effects of these soft collisions can result: (1) the two colliding particles can stick together, forming loosely-bound larger particles; (2) if one of the particles is already a loosely-bound conglomerate, the collision can break it apart into smaller components; or (3) the collisions can be approximately elastic, with the two participating particles emerging from the collision along slightly altered paths. All of these effects can be combined under the term *viscosity*, so called because of the similarities between these effects and that of the motions of a viscous liquid. The analogy becomes even more precise when one considers that the motions of the individual particles cannot be tracked, but only the resulting changes in the appearance of the ring.

The most apparent result of gravitational resonances is the formation of spiral *density waves*, in which the outwardly nudged ring particles at a resonance position crowd closer to ring particles at a slightly larger orbital radius. The regions of denser ring population thus created exert a gravitational force on nearby ring particles, and the net effect is to propagate the disturbance in a direction outward from the resonance radius in a tightly wound spiral that gets weaker as the density wave gets farther and farther from the resonance radius. The distance between the density crests and troughs and the contrast in particle number density between the peaks and valleys is dependent on several factors, including the *mass density* (the mass per unit area of the ring material, integrated through the thickness of the ring) of the ring at the resonance radius, the mass and distance of the satellite causing the density wave, and the amount of variation in ring particle velocities near the resonance position. Thus, careful measurements of density waves can lead to a determination of the mass of the perturbing satellite, the mass per unit area within that portion of the ring, and an estimate of the distribution of ring particle velocities near the resonance point.

If the satellite responsible for the gravitational perturbations of the ring is inclined to the ring plane, a different kind of effect is noted. In this situation, the perturbed ring particles move above and below the ring plane, and a *bending wave* is created. Bending waves involve vertical oscillations that have periods slightly longer than the orbital period (the oscillation periods of ring particles in density waves are slightly shorter than their corresponding orbital periods). In bending waves, the spiral wave ("corrugation") propagates inward toward the planet from the resonance radius, opposite the propagation direction of density waves. Of the close-in satellites, only Mimas, with an inclination of 1.572°, has a combined mass, distance, and tilt sufficient to give rise to identifiable bending waves. Bending waves are thought to be the main contributor (or at least one of the primary contributors) to the visibility of the main rings when they are viewed edge-on. Examples of a density wave and a bending wave, both due to the 5:3 resonance with Mimas in Saturn's A ring, were shown earlier in Figure 1.5.

Other effects from gravitational resonances are also apparent in Saturn's rings. The 7:6 resonance with Janus is responsible for confining ring particles at the outer edge of the A ring (compare periods in Table 9.1). The outer edge of the B ring is near the 2:1 resonance with Mimas. These two ring boundaries are consequently seven-lobed and two-lobed, respectively. The two-lobed shape of the outer B-ring edge is clearly seen in images from Voyager 2 (Figure 9.9), where an offset of about 50 km is seen in the radius of the outer edge of the B ring at two different azimuths. The gap at the outer edge of the B ring in Figure 9.9 is called the Huygens gap. Within the gap is (at least) one narrow elliptical ring, possibly presaging an as-yet-undiscovered satellite within the Huygens gap. Note also in the figure that fine structure in the outer B ring does not match well between the images, although the structure in the Cassini Division to the right in the figure does fit well. The cause of the mismatched fine structure is unknown. The smallest features in the figure are about 7 km across.

Embedded satellites can also clear gaps within rings. In fact, it is suspected that all clear gaps within the ring system are the result of particle sweeping by a satellite within the gap (see Section 9.6). Nevertheless, prior to Cassini's arrival at Saturn, only one such embedded satellite had been discovered in the Saturn ring system, namely Pan, which orbits Saturn in the center of the Encke gap in the outer half of the A ring. Pan also disturbs the material near the Encke gap, causing waves along its edges with a particular length, much like water flowing over a rock in a stream. These waves get longer in material further from the satellite—causing ring material to bunch in radial structures called "moonlet wakes". Particles near the inner edge of the Encke gap circle Saturn in less time than Pan; the edge waves and wakes at the inner boundary of the Encke gap thus precede Pan in its orbit. Edge waves and wakes at the outer boundary trail Pan in its orbit. In fact, the wakes were seen before Pan itself (see Figure 9.10) and led to its discovery in Voyager images nearly 10 years after the Voyager encounters [31]. The discovery of at least one other embedded satellite has resulted from the observation of edge waves in the Keeler gap near the outer edge of the A ring (see Section 10.4).

Voyager observations of Saturn's F ring and the discovery of Prometheus and

**Figure 9.9.** This composite image of four views of the outer edge of the B ring was assembled from images shuttered by Voyager 2. At the left is the bright B ring; at the right is the darker Cassini Division. Separating the two is a gap of variable width known as the Huygens gap. A narrow eccentric ringlet is seen within the gap. Resolution in the image is about 7 km. (260-1473)

Pandora on either side of it led scientists to believe that the gravitational influence of these two satellites somehow confined and maintained the particle population of the narrow F ring. They have long been known as the "F-ring shepherds." A similar mechanism was proposed to explain the existence of narrow rings at Uranus and Neptune (see Chapters 7 and 8). Voyager and subsequent Earth-based observations of the F ring now seem to indicate substantial evolution of the F ring in both its character and its width. That in turn has led many ring scientists to question whether Prometheus and Pandora exert any appreciable forces on the F ring that confine its natural spreading; they may not deserve the appellation of F-ring shepherds.

**Figure 9.10.** This view of the 325-km wide Encke gap has been stretched horizontally to emphasize the edge waves along the right (outer) edge. The waves are due to Pan, whose existence was predicted on the basis of these edge waves but not identified in Voyager images until many years later. (Voyager 2 image FDS 43993.50, from Cuzzi and Scargle [31])

Not all of the structure observed in the rings of Saturn is traceable to the types of interactions discussed above. In fact, it is fair to say that there is far more structure in Saturn's rings that remains unexplained than there is that we fully comprehend. Examples of features not yet understood are the semi-regular banded structure of the C ring, much of the fine-scale radial structure seen throughout the A and B rings, and the irregular structure seen in the outer 1,000 km of the B ring. Some are possibly due to gravitational interaction with as-yet-undiscovered satellites, some may be a consequence of high particle volume density and frequent collisions between those particles, and some could be a consequence of frequent random impacts of interlopers (dust grains, meteoroids, etc.). Some of the structure persists for long periods of time and some is time-variable, but there is much still to be learned before we can claim to have a relatively complete understanding of Saturn's ring system.

## 9.6 GAPS AND NON-CIRCULAR FEATURES IN SATURN'S RINGS

In the previous section, we mentioned the narrow gap at the outer edge of the B ring. There is at least one ringlet in that gap. As can be seen from Figure 9.9, the ringlet is eccentric. Multiple partial rings also exist in the Encke gap in the outer A ring, where Saturn's satellite, Pan, was discovered. In some ways, these partial rings have characteristics reminiscent of Neptune's ring arcs or the denser portions of the F ring. It is

**Table 9.2.** Known gaps within Saturn's rings.

| Ring region | Name of gap | Radius (km) | Radius ($R_S$)* | Width (km) | Notes |
|---|---|---|---|---|---|
| C ring | Unnamed | 74,900 | 1.241 | ~70 | Several features |
| | Colombo | 77,800 | 1.290 | 184 | Eccentric ringlet |
| | Maxwell | 87,500 | 1.450 | 270 | Eccentric ringlet |
| | Unnamed | 88,700 | 1.470 | < 10 | Opaque ringlet |
| | Unnamed | 90,200 | 1.495 | 20 | Opaque ringlet |
| Cassini Division | Huygens | 117,820 | 1.953 | 285–440 | Eccentric ringlet |
| | Unnamed | 118,200 | 1.959 | 38 | |
| | Unnamed | 118,300 | 1.960 | 28 | |
| | Unnamed | 118,600 | 1.966 | 40 | |
| | Unnamed | 119,000 | 1.972 | 42 | |
| | Unnamed | 119,900 | 1.988 | 246 | Opaque ringlet |
| A ring | Encke | 133,570 | 2.214 | 325 | Several ringlets |
| | Keeler | 136,530 | 2.263 | ~35 | |

$1R_S = 1$ radius of Saturn = 60,330 km.

possible that narrow eccentric and/or partial rings and small satellites are characteristic of the other clear gaps in Saturn's rings. Again, the Cassini Orbiter holds the key to a number of potential discoveries and better understanding of these gap features.

There are very few clear gaps in Saturn's rings. Already mentioned are the Encke and Keeler gaps in the A ring and the Huygens gap at the inner edge of the Cassini Division. Additionally, the Maxwell gap in the C ring is a major clear gap within the rings of Saturn. Table 9.2 lists the clear gaps known prior to the arrival at Saturn of the Cassini spacecraft. The table has been adapted from table V of Cuzzi *et al.* [32]. Note that many of these gaps have one or more narrow, often discontinuous ringlets within them. It also seems likely that each gap has one or more moonlets within it that are in part responsible for clearing the gap. Each may additionally have narrow ringlets whose source may well be debris from the embedded moonlet. By analogy with the Encke gap, edge waves, now known to be gravitational wakes of the embedded moonlet Pan, may be present in each of the gaps.

## 9.7 RARIFIED GASES IN THE VICINITY OF SATURN'S RINGS

It has long been known that water ($H_2O$) ice is the primary constituent of the rings of Saturn. Various mechanisms should produce a tenuous atmosphere near the rings that reflects this composition. The primary constituents of such an atmosphere were expected to be hydrogen (H or $H_2$) and hydroxyl (OH). The most direct way to measure such an atmosphere would be with instruments like those on the Cassini Orbiter designed to sense this very thin atmosphere as the spacecraft moved through it. Indeed, that is exactly what happened during the Saturn Orbit Insertion time period on July 1, 2004 (see Section 10.3)

The most extensive pre-Cassini observations of a ring atmosphere were by the ultraviolet spectrometers aboard Voyagers 1 and 2 [33]. The only component measured was Lyman-alpha radiation from atomic hydrogen (H). Because the solar system is full of hydrogen, the radiation from Saturn's rings was a small fraction of the total hydrogen observed, and the estimate of about 360 Rayleighs from Saturn's rings was more or less independent of whether the observation was of the illuminated rings or the unlit face of the rings. It was also independent of the tilt of the rings to the Sun, which was about 3.6° for Voyager 1 and 8° for Voyager 2. The hydrogen emission intensity dropped off very slowly with altitude above the rings, decreasing by a factor of 2.72 over about 6,000 km and was not observed at distances from Saturn greater than the outer edge of the A ring. The source of the hydrogen may be high-velocity meteoroid impacts on the ring particles [34], the only as-yet-proposed mechanism with high enough temperatures to produce—from the rings themselves—the amounts of hydrogen observed.

A source external to the rings may also be a possible source for the atomic hydrogen. It has been suggested that molecular hydrogen ($H_2$) from the outer portions of Saturn's atmosphere might be *dissociated* by electrons in Saturn's radiation belts into hydrogen atoms of sufficiently high energy to escape the planet's gravity [35]. The escaping hydrogen atoms would then lose energy as they encountered the cold ring particles and remain trapped near the rings. However, it is certainly conceivable that the observed hydrogen is simply part of the extended hydrogen atmosphere of Saturn that surrounds the planet and its rings, and that these early measurements mistakenly identified it as being a part of a ring atmosphere or affected by the rings.

A more complete understanding of the ring atmosphere will require Cassini data, both the *in-situ* data mentioned above and additional remote sensing, and supporting theoretical analyses, most of which are still unavailable as this book goes to press.

## 9.8 RING-PARTICLE PROPERTIES

The beginning of our understanding of the sizes of the ring particles came as a result of the discovery in 1973 that Saturn's rings are strong radar reflectors [36]. Unless the ring particles were at least as large as the radar wavelength (12.6 cm), there would have been no reflected radar signal detected. The rings were expected to be transparent to radar on the basis of failure to observe any passive radio signals from Saturn's rings prior to that date. The transmitted radar signal was polarized, but the reflected signal was largely depolarized. That meant that the ring particles reflecting the radar signal were moderately irregular [37]. The same radar measurements showed a reflectivity for the rings that was inconsistent with silicate (rocky) composition for the rings. Such high reflectivity was consistent only with metallic composition (which was highly unlikely) or with relatively pure water-ice composition.

There is evidence in the Voyager and Earth-based data for Saturn ring particle sizes ranging from dust-sized particles of a micrometer or less in radius to embedded satellites akin to Pan with a radius of about 10 km. The primary pre-Cassini constraints on ring particle size are provided by the radio occultation and stellar occultation data, the observed light-scattering characteristics of the individual rings, analysis of data indicating particle hits on the Voyager spacecraft, and theoretical considerations.

The radio occultation data on Voyager 1 [38] will be discussed in detail later (Section 9.12). The difference in transparency of the rings at the 13-cm and 3.6-cm radio wavelengths provides a measure of the relative abundance of particles in the size range from 1- to 4-cm radius. That relative transparency varies greatly with location in the rings.

There is almost no difference between the X-band and S-band radio transparencies of the Cassini Division; this ring segment must have a large population of ring particles greater than 10 cm in radius. The C ring, on the other hand, shows significant differences and must therefore have many particles in the 1- to 4-cm size range. The B-ring noise levels are higher, and except for its innermost part, little useful data on particle sizes was obtained. Interpretation of radio occultation data, especially those of the B ring, was also made more difficult by the low tilt angle of the rings as viewed from the Earth in 1980. It appears that the greatest differences between S- and X-band transparencies occur in the outer A ring, implying that it possesses an abundance of particles between 1 and 4 cm in radius. It is also noteworthy that the greatest differences between optical and radio depths also occur in the outer half of the A ring, which must therefore also have a large fraction of particles under a centimeter in radius. It is possible that at least part of the difference is due to the low tilt angle of the rings during the radio occultation measurements, combined with the likelihood that larger particles that block the radio signal occupy a smaller fraction of the total vertical thickness of the outer A ring.

The D, G, and E rings were detected in neither the radio nor the stellar occultation experiments. The implication is that the D, G, and E rings are dominated by dust-sized particles. The core of the F ring was detected at X-band, S-band, and optical wavelengths. The radio detections implied a much narrower ring than seen in visible-light images. The F-ring core must therefore contain large particles, but the more extended F-ring structure mainly comprises dust-sized particles, perhaps with a small population of larger (but still sub-centimeter) particles. Of course, the data do rule out even larger bodies (moonlets) widely separated in ring longitude.

Now, lest our readers get the false impression that the particle size distribution in the rings is well determined, consider a recent paper by French and Nicholson [39], based on a re-analysis of the stellar occultation data by the Voyager photopolarimeter at three wavelengths: 3.9, 2.1, and 0.9 micrometers. In this analysis, instead of analyzing the directly transmitted starlight, they processed the data to calculate that portion of the signal due to the directly transmitted light and subtracted it from the total signal. The remainder was due to light scattered from the ring particles; the sharpness of the angular distribution is set by the largest particles and the breadth of the angular distribution reflects the abundance of smaller, centimeter-sized particles. By assuming uniformity in particle size distribution across major ring regions, they found characteristic particle sizes in the A and C rings that are fairly consistent with those derived from the radio occultation data. Their analysis of the B-ring data seems to indicate minimum characteristic diameters of about 30 cm and maximum characteristic diameters of about 20 m. Hopefully, Cassini data will provide better information on particle size distributions within all of Saturn's rings.

Both Earth-based and Voyager measurements of color differences within the main rings make it clear that there are at least small amounts of *contaminant* mixed in with the water ice, especially in (but not limited to) the C ring and the Cassini Division. However, it doesn't take much contaminant to cause the observed coloration, certainly no more than 10% of the ring material and perhaps as low as 1% [40].

The D ring has not been seen in backscattered light (i.e., with the observer between the Sun and the D ring). It was imaged by both Voyager 1 and Voyager 2 in forward-scattered light (with the D ring between the observer and the Sun). The Voyager 2 image is reproduced in Figure 9.11. The D ring is the only one of the tenuous rings of Saturn (D, G, and E rings) with narrow ring structure, reminiscent of the Uranus dust disk observed by Voyager 2. The source of the radial structure is unknown. From the greatly enhanced visibility which occurs at large phase angles when the ring particles are comparable in size with the wavelength of the light, the particles are thought to be about a micrometer in radius.

The F ring is still somewhat of an enigma. It may be a recent, unusually large collision product [41]. Detection of its extended width in stellar occultation measurements and of only its core in radio occultation measurements seems to imply that there are few centimeter-sized and larger particles away from its central core, but that larger particles exist within that core. The F ring is known to be highly variable on a range of timescales [42]. In particular, the brighter "knots" seen in Voyager 1 imaging are not correlated with those seen by Voyager 2, and one bright region in a series of Voyager 2 images generated several new clumps which orbited at different rates from the bright region itself [43]. Information on its composition comes primarily from visual imaging; its redder color relative to the A ring suggests a larger fraction of silicate material, although that conclusion is somewhat tentative. A better understanding of the F ring needs repeated detailed measurements by an orbital spacecraft like Cassini (see Chapter 10).

The G ring does not have the apparent fine structure of the F ring and it is much wider and more diffuse in character (see Figure 9.12). Showalter and Cuzzi initially concluded that the G ring was composed of a broad (5,000 km) band of fine dust grains, which might contain a narrower core of large objects, as had earlier been suggested by Van Allen [44], which served as the source of the G-ring material. In the past, the closest counterpart to the G ring was thought to be the E ring. Several more recent (but still pre-Cassini) papers challenge this view. The G-ring material is confined to a narrow vertical layer across its entire width, unlike the vertically extended E ring [45]. It also has a reddish hue, in contrast to the distinct bluish color of the E ring [46]. Impacts on the spacecraft during passage through the edges of the G ring are also incompatible with the narrow size distribution which characterizes the E ring [47]. These results imply that there are larger ring particles across the entire G-ring width and that their composition is distinctly different from the water-ice grains that make up the E ring.

**Figure 9.11.** Saturn's D ring is best viewed in forward-scattered light; like dust on an automobile windshield is much more visible when driving toward the Sun. This image is from Voyager 2 from a distance of 195,400 km. The phase angle (angle between the Sun and the observer as seen from the target) is 166°, or only 14° from the direction of the Sun. The dark band at the right edge of the image is Saturn's shadow across the D ring. (PIA01388)

The E ring extends from the orbit of Mimas (which orbits at a mean distance of 185,520 km) to a radial distance of at least 483,000 km from Saturn's center and perhaps nearly to the orbit of Titan (orbital radius of 1,221,830 km), as derived from Cassini orbiter data and depicted in Figure 2.14. The thickness of the E ring is about 1,000 km near the orbit of Enceladus (orbital radius of 238,020 km), where it is also brightest. The physical thickness increases to more than 15,000 km in its outer portions, but its optical thickness decreases with distance beyond the orbit of Enceladus until it becomes essentially optically undetectable. The particle size, as determined from analysis of particle impacts with the two Voyager spacecraft and from

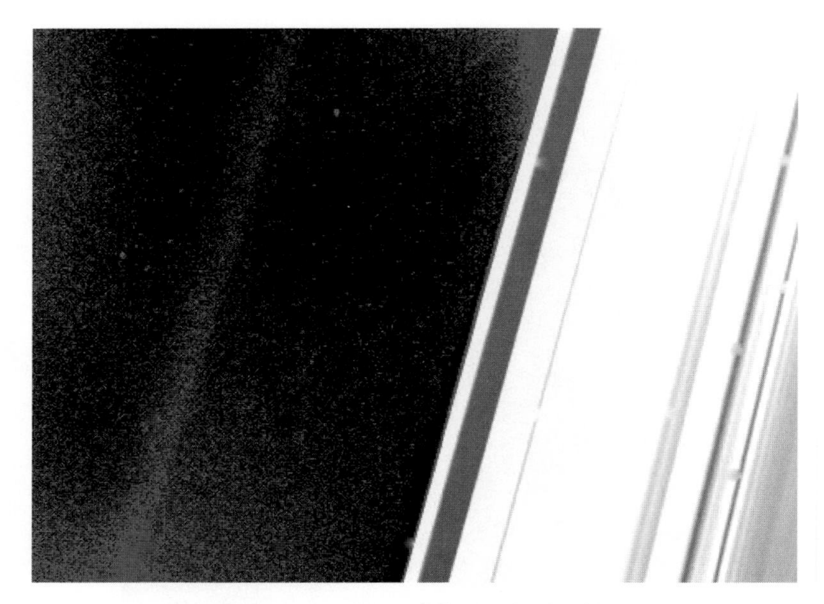

**Figure 9.12.** Saturn's G ring is seen as a faint, diffuse stripe in this long-exposure image taken by Voyager 2's narrow-angle camera from a range of 305,000 km on August 26, 1981. The overexposed F and A rings are seen in the right-hand part of the image. The white dots seen in and near the F and A rings are incompletely removed reseau marks on the camera face that are used for geometric reconstruction of the image. (PIA01964)

Earth-based observations, is very narrowly centered near a radius of one micrometer. The particle composition is probably water ice, and even prior to Cassini, it was suspected that the source of those water-ice grains was Enceladus itself, although the mechanism for generation of those grains was not well understood. This is discussed in more detail in Chapter 10.

## 9.9 REFLECTIVE PROPERTIES OF SATURN'S RINGS

The brightness of a ring or a portion of a ring is dependent on both the total reflectivity (*albedo*) of the particles within that ring or ring portion and on the directional distribution of the reflected radiation (*phase function*). The phase function is the variation in brightness of the target as the phase angle varies between 0° (back-scattered light) and 180° (forward-scattered light). Sometimes the term *scattering angle* is utilized. The scattering angle is simply the supplement of the phase angle; or, in other words, the scattering angle is the difference between the phase angle and 180°. In mathematical terms,

$$\Theta = 180^\circ - \alpha,$$

where $\Theta$ is the scattering angle and $\alpha$ is the phase angle.

The phase function for the bulk of the A and B rings peaks strongly at 0° phase angle. This characteristic is known as the *opposition effect*. Saturn's A and B rings have a much sharper opposition surge than any other natural objects in the solar system. Cassini observations have brought a much better understanding of this effect (Chapter 10).

In marked contrast with the strongly back-scattering A and B rings, the D, G, and E rings are strongly forward-scattering. This behavior is typical for rings which are dominated by particles in a size range that is comparable with the wavelength of the light incident on the rings (approximately a micrometer in radius).

## 9.10 THERMAL PROPERTIES OF SATURN'S RINGS

The temperature of any given ring particle is the result of a balance between energy absorbed by the particle and energy re-radiated by the particle. The energy absorbed by a ring particle is dependent on several factors: the amount of sunlight that is incident on the ring particle (either direct, reflected, or scattered), the amount of thermal radiation received from other ring particles or from Saturn itself, the total reflectivity (*bolometric Bond albedo*, integrated over all directions and over all wavelengths) of the ring particle, the spin rate of the ring particle, and the size of the particle.

The amount of sunlight that reaches a particle is dependent on the tilt angle of the rings (from 0° to 26.73°) and on the optical thickness of the ring region. As one might suspect, the greater the tilt angle, the more sunlight reaches any given ring particle, at least for optically thick rings like the A and B rings. The absolute temperature of the A and B rings is nearly twice as high (90 K to 100 K) when the ring tilt angle is at its highest compared with that (~50 K) when the rings are edge-on to the Sun. Contrast that with the temperatures of the C ring and Cassini Division, where measured temperatures are between 80 K and 90 K and almost no temperature variation with ring tilt angle is observed [48]. The probable cause of the uniform temperatures with tilt angle for the C ring and Cassini Division is the low optical thickness of these rings, which probably means that interparticle effects are small.

The C ring and Cassini Division particles also have lower reflectivity at visible wavelengths than do the A- and B-ring particles and therefore absorb more of the radiation incident on them. This alone, however, is insufficient to explain their higher measured temperatures. In addition, the temperatures must be higher (as observed from Earth [49]) due to the presence of particles that are large enough and slowly rotating enough that little heat is conducted from their illuminated sides to their unilluminated sides. That conclusion seems to be borne out by Cassini observations; moreover, Cassini data may be starting to show us variations in the spin rate of particles from place to place (Chapter 10).

The C ring and Cassini Division have essentially the same measured temperatures whether observed from the illuminated or unilluminated face of the rings. The dark faces of the A and B rings are considerably lower (about 56 K [50]) than their illuminated faces, except at very low tilt angles. Particles on the unilluminated face generally transition to the illuminated face at least once each orbit around the planet. The lower temperatures imply that the ring particles can cool quickly and heat quickly in response to changing energy input (i.e., they are composed of material with low *thermal inertia*) and that they also have highly insulating surfaces that do not allow heat to be conducted clear around or through the interior of the particles. This is also consistent with the observation of substantially lower temperatures in the A, B, and C rings after passage through Saturn's shadow than before shadow entry [51].

## 9.11 EARTH-BASED OBSERVATIONS OF SATURN'S RINGS SINCE 1981

There are three primary types of observations of Saturn's rings that have provided additional information on the ring system since the Voyager encounters in 1980 and 1981. These include an occultation of the star 28 Sagittarii by Saturn and its rings in 1989, observations at the time of the ring plane crossings of the Earth and the Sun in 1995 and 1996, and radar observations of the rings from the Arecibo Observatory in Puerto Rico from 1999 onward. In addition, there was an extended series of color observations of Saturn's rings using the Hubble Space Telescope and spectroscopic observations of the rings using the Infrared Telescope Facility atop Mauna Kea on the island of Hawaii. There have also been improved passive radio interferometric measurements of Saturn's rings.

Analysis of the 28-Sagittarii occultation is presented in three papers [52]. The apparent size of the star at the distance of Saturn was about 20 km, which represents a lower limit on the size of radial features that could be detected in the rings. Several occultations by Saturn's rings of dimmer stars were also observed from the Hubble Space Telescope, providing some information on both radial and azimuthal variations in the rings [53].

The motion of Saturn around the Sun causes the Saturn ring plane to cross the Sun once every half-orbit (about 15 years). Depending on the position of the Earth in its orbit around the Sun, the ring plane crosses through the Earth either once or three times near Saturn's equinox, when the ring plane is edge-on to the Sun. As of this writing, the most recent ring-plane crossings of Sun and Earth occurred in 1995 and 1996. Among other revelations, observations near the ring-plane crossing provided new insights into the time-variable nature of clumps in the F ring [54], a better definition of particle size distribution in the G ring [55], and discovery of the orbital changes of Saturn satellites Pandora and Prometheus [56].

Relatively high-resolution radar imaging of Saturn's rings was conducted annually, beginning in 1999. These "images" (see Figure 9.13) were constructed by plotting the returned radar signal delay time (distance) versus the signal frequency, which is spread out as a result of the orbital motion of the rings, due to an effect known as *Doppler shift* [57]. The results of several years of observations are discussed by Nicholson *et al.* [58]. The azimuthal asymmetry of the A ring (discussed in Section 9.4 and Figure 9.7) is even more apparent at radio wavelengths than it is at visible wavelengths, and the positions of maxima and minima are essentially identical in

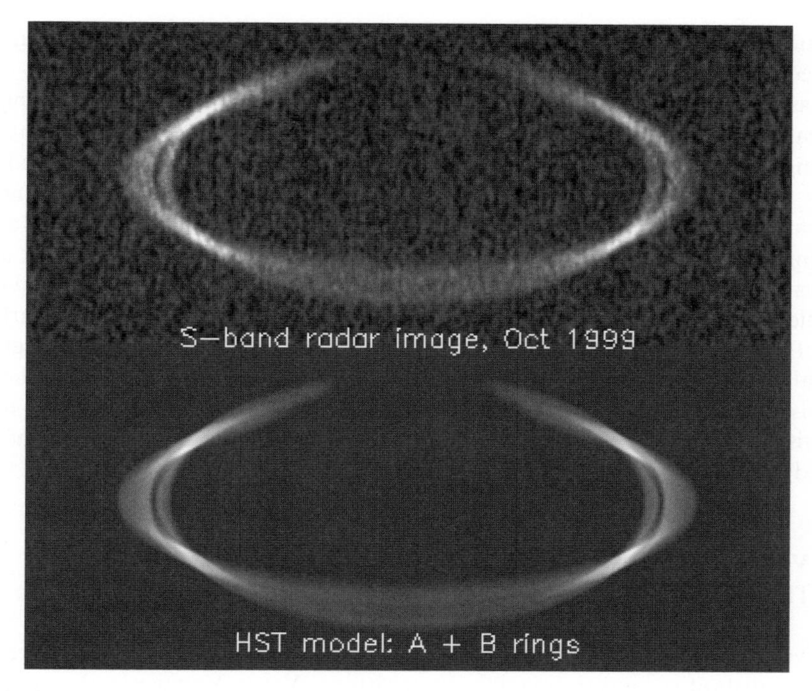


**Figure 9.13.** The image at the top is a sum of 5 days of radar data in 1999. The image is constructed by displaying data in time-of-flight increments, with time delay increasing from bottom to top. The left-to-right dimension is produced by the Doppler-shifted frequencies of the returned radar signal, with frequency increasing from left to right. Note that—although the A ring is exterior to the B ring (the C ring was not detected) at the top and bottom of the image—since the B-ring particles move more rapidly than the A-ring particles, they have greater Doppler shifts and are exterior to the A ring at the left and right in the images. This leads to four bright "crossover" regions on either side where the signals from the two rings add together. Note that it also results in a left-to-right reversal of the rings relative to visible-light images. A pronounced azimuthal asymmetry can be seen: the rings are brighter on the far quadrant on the receding (left-hand) ansa and on the near quadrant of the approaching (right-hand) ansa. This "delay-Doppler" image of Saturn's rings is at an S-band radar frequency of 2,380 MHz (12.6 cm) and is compared with a model image constructed by reprojecting a pair of Hubble Space Telescope images taken at visible wavelengths. The effective spatial resolution of the radar image is 2,000 km (left to right) by 15,000 km (top to bottom). (Image from Nicholson *et al.* [58])

both data sets. The radar data also show a similar, but less pronounced effect in the B ring.

Hubble Space Telescope observations [59] succeeded in imaging radial spokes within the B ring similar to those seen in Voyager images (Chapter 2 and Section 9.13) and in determining the size of the particles making up the spokes. However, as the ring tilt angle became even higher, the B-ring spokes vanished. This has been puzzling (see Chapter 10). High-spectral-resolution data from the Infrared Telescope Facility [60] confirmed that the primary chemical constituent of the rings is water and also provided evidence that will help to identify several minor constituents.

Passive radio-wavelength measurements of the temperature of Saturn's rings have been carried out by several observers [61]. The measured temperature at 1-mm wavelength and longer is significantly lower than that measured at infrared wavelengths of a few micrometers. These lower measured temperatures are not real temperatures, but are the consequence of the low *emissivity* of ring particles at radio wavelengths. Emissivity is a measure of the efficiency with which a body at a given temperature emits infrared and radio thermal radiation. Generally, the emissivity of a low-reflectivity body is high; conversely, as the reflectivity (albedo) of a surface or array of particles increases, the emissivity decreases. At mid-infrared wavelengths, the physical temperature and measured brightness temperature are essentially the same; the emissivity at those wavelengths is close to 1.0. At wavelengths of a few mm the emissivity and brightness temperature of Saturn's rings drops to about 10% of that at infrared wavelengths.

## 9.12 VOYAGER RADIO OCCULTATION DATA AND THEIR INTERPRETATION

In the view of the general public, the radio observations of Saturn's rings generally came in a distant second to imaging observations. Yet in Voyager's observations, it was the radio occultation observations which provided the highest radial resolution data on the rings, which gave the most direct information on ring particle sizes, and which seemed to confirm the suspected nature of the A-ring azimuthal asymmetry. Details of the Voyager 1 observation technique and theory are given by Eshleman *et al.* [62] and Marouf *et al.* [63].

Radio occultation observations of the C ring and the Cassini Division revealed differences between S-band and X-band transparencies that are much smaller than those in the A ring. This is consistent with the observation that optical opacity and (adjusted) radio opacity are essentially identical in these regions. The adjustment is a reduction (by a factor of 2) in the calculated radio depths (opacities), related to the *coherency* [64] of the radio beam; starlight used in the optical occultation measurements is non-coherent. A detailed discussion of ring particle size distributions from Voyager 1's radio occultation experiment is given by Marouf *et al.* [65].

Within the rings, the radial resolution of the radio occultation measurements is approximately 15 km, but for those areas where the Earth-received signal is strong enough, phase matching can be used to remove the diffracted part of the signal (which originates from portions of the ring away from the center of the beam), thereby improving the radial resolution to better than 1 km [66]. This higher radial resolution was not possible in the denser inner B ring, where the opacity was significantly larger and the signal strength correspondingly weaker, but resolutions on the order of a few kilometers were achieved.

The rings were sampled by the Voyager 1 radio occultation experiment over a radial range that spanned from inside the C ring to beyond the F ring. The ring tilt angle (as seen from the radio receivers on Earth) was 5.9° at the time of the experiment. Practically, this small tilt angle limited the sensitivity of the experiment to radio depths from about 0.005 to about 1.0. In the outer half of the B ring, derived X-band radio depths exceeded in several places a value of 1.0, very near the upper limit of sensitivity. For that reason, the Voyager 1 outer B-ring radio opacities are not as reliable as those for the inner B ring and for the other rings; the S-band sensitivity in the outer B ring was much worse.

The radio occultation data have also been processed in such a way as to provide some information on sub-centimeter-sized particles in the A and C rings and Cassini Division and the distribution in particle sizes in those same rings for particles in the 1- to 10-meter radius size range [67]. The sharp definition of the density waves from radio science data, especially in the A ring, have provided the best values for ring density per unit area and for the masses of the perturbing satellites which give rise to the density waves. In short, the radio occultation experiment produced a wealth of data on Saturn's rings that not only helped to define the rings but provided a strong basis for the follow-on studies of Saturn's ring system by the Cassini Orbiter.

## 9.13 INTERACTIONS BETWEEN SATURN'S RINGS AND MAGNETIC FIELD

Gravitational forces by seen and unseen satellites and by Saturn itself probably generate the majority of Saturn's ring features. Saturn's gravity, with the help of meteoroids from outside the Saturn system and mutual collisions between the ring particles themselves, is responsible for the breakup of larger bodies that approach within the Roche limit (see Chapter 1). The oblateness (polar flattening) of the planet also keeps the rings and the satellites which interact gravitationally with the rings close to the equatorial plane, thereby increasing the opportunity for particle–particle collisions. Particle collisions in turn work toward the circularization of ring-particle orbits and the viscosity effects discussed briefly in Section 9.5. It is also the gravity of Saturn that creates Keplerian shear (Section 9.3) as ring particles closer to Saturn orbit at faster speeds than those farther from the planet. Nevertheless, there are observed phenomena within the rings that seem to defy the behavior dictated by gravitational forces alone, and it is those we discuss in this section.

Pioneer 11 was the first spacecraft to reach Saturn. The spatial density of ions and electrons was measured by Pioneer 11 as it probed to within a range from Saturn's center that corresponded to the middle of the C ring. Inward of about 8 Saturn radii (i.e., just interior to the orbit of Rhea) Pioneer 11 found a plasma that increased in density as one approached the planet and co-rotated with the planet. The charged particles in that "cold" plasma were essentially frozen into the magnetic field of the planet, circling the planet around an axis parallel to Saturn's rotation axis in the same amount of time as the planet's rotation, approximately 10 hr 39.4 min. While these charged particles were constrained to revolve around Saturn at the same rate as Saturn's rotation and were further constrained not to move radially inward or outward, their motion in a north–south direction (along magnetic field lines) was less restricted.

As Pioneer 11 reached the radial distance of the A and B rings, scientists noted that there was a sudden decrease in the spatial density of these plasma particles. Apparently the north–south motion of the plasma resulted in its absorption by ring particles, creating a plasma "vacuum" above and below the rings. Pioneer plasma scientists also noted somewhat diminished plasma densities near radial distances of about 140,000 and 170,000 km from Saturn's center [68]. These were the radial distances of the F and G rings. Although spin-scan images of the F ring were obtained by Pioneer 11 [69], thus verifying the existence of ring material near 140,000 km, the G ring was not imaged until Voyager 1 obtained an image in 1980. The only evidence for its existence prior to that image was the observations of depleted plasma density at that distance.

In spite of the sharp decrease in plasma density inward of the outer edge of the A ring, there are detectable levels of protons and electrons within that region. It is suspected that these protons and electrons come from the impingement of cosmic rays on ring material [70], a conjecture that is consistent with the presence of water ice as the primary chemical constituent of the rings.

B-ring spokes (Figure 9.14) are believed to form as meteoroids strike and shatter B-ring particles, spreading the debris across tens of thousands of kilometers [71]. These collisions seem to prefer the shadow region of the rings, which is the region where the highest velocity impacts would occur from Sun-orbiting meteoroids with high eccentricities and inclinations [72]. By some mechanism, possibly the impact of magnetospheric plasma or by the action of sunlight on these tiny particles, they apparently obtain an electrical charge large enough that the dust grains in the spokes are partially frozen in the magnetic field and thus revolve around Saturn without losing their shape for a time. It is interesting to note that ring particles about $\frac{3}{4}$ of the way out in the B ring have Keplerian velocities that equal the co-rotational velocities at that distance, so the electromagnetic forces needed to perturb B-ring spoke particles to co-rotation are relatively small. The spokes apparently dissipate and disappear as the particles lose their charge (perhaps as they pass through the ring plane twice each orbit) or adhere to larger particles, and assume normal Keplerian orbits.

The correspondence of the position in the outer B ring where Keplerian velocity equals the co-rotation velocity is known as a *Lorentz resonance*. Tiny ring particles slightly inward or outward of the Lorentz resonance radius experience small forces that tend to push them away from the resonance radius. Over time, that should deplete that region of the B ring of dust-sized particles, were they not replenished by other processes. Lorentz resonances also occur at other small-number fractional relationships between the ring orbital periods and Keplerian orbital periods. Lorentz forces play a major role in the structure of Jupiter's rings (see Chapter 6).

Another important electromagnetic effect within Saturn's rings is known as Poynting–Robertson drag. Technically, this is an effect that is due to the absorption and re-emission of sunlight by dust-sized particles, but because light is an electromagnetic radiation, Poynting–Robertson drag is included in this section. Tiny ring particles orbiting Saturn absorb sunlight from one direction and re-emit that energy in

**Figure 9.14.** This image of spokes in the B ring was obtained from Voyager 2's high-resolution camera from a distance of 4 million km on August 22, 1981. The image was taken at low phase angle, and the fine-grained spokes appear dark against the B ring. At high phase angles, the spokes appear bright against a darker B ring, both on the illuminated and non-illuminated faces of the rings. (PIA02275)

a slightly different direction, resulting in a net force that, over time, causes the particle to spiral inward toward the planet. Eventually the particle will enter Saturn's atmosphere and be lost to the ring system. This process becomes more and more efficient at smaller and smaller ranges from the planet and for tinier and tinier particles, and it is an important factor in determining the effective lifetime of Saturn's (or any other planet's) ring system.

## 9.14 SUMMARY AND MAJOR UNANSWERED QUESTIONS BEFORE CASSINI

The purpose of this chapter has been to set the stage for the spectacular data on Saturn's rings returned by the Cassini Orbiter. In Chapter 10 we report on some of the preliminary findings of this important international space mission. Hopefully, the data we include in this book will whet the appetites of interested readers sufficiently to cause them to search for the later results that will already have appeared or will shortly appear that further expand on our understanding of the Saturn ring system.

The rings of Saturn have been observed for centuries; none of the other ring systems was even known to exist prior to 1977. They have progressed from ignominy to passing familiarity, primarily as a consequence of the Voyager 1 and 2 encounters in 1980 and 1981. By the time the Cassini ring data are fully absorbed and digested, they will be old friends, but like the best of our friends, there will remain many things about them that are still strange and seemingly idiosyncratic. In part, it is the difficulty associated with delving into their complexity, and the thrill of recognition that they are slowly but surely helping us to understand the ways of Nature, that endear them to us.

The Saturnian rings are evolving on a number of timescales ranging from days to eons. Those changes are not apparent to observers with small to intermediate telescopes, where the bejeweled planet is sometimes seen to have atmospheric storms that come and go, but appears to be surrounded by an unchanging set of rings. But, to a trained observer using Earth-orbiting telescopes or large ground-based telescopes equipped with resolution-enhancing adaptive optics or sophisticated robotic spacecraft that carry our eyes and ears to Saturn, the changes in the rings are unmistakable. It is the understanding of those changes that will eventually enable space scientists to extend that understanding to the asteroid belt, to the trans-Neptunian Kuiper Belt, to protoplanetary disks around other Milky Way stars, to spiral structure in the Milky Way Galaxy and other galaxies, and possibly even to clusters of galaxies.

Now let us examine briefly each of Saturn's rings, summarizing some of the known and unknown characteristics of each.

We know the radial location of the outer edge of the D ring, but its inner edge is indistinct, and the forces which maintain those boundaries are poorly understood. We see radial structure within the D ring, but we do not know what causes that structure, and Voyager and Earth-based data are insufficient to show us changes in that structure. Dust-sized particles likely dominate its population, judging by its brightness in forward scattering, but we do not know the thickness of the ring or whether its particles are affected by Lorentz resonance. The mass of the ring is also unknown.

The C-ring inner and outer boundaries are known, but the processes which give rise to the sharp drop in ring brightness and optical depth in the B-ring to C-ring transition are unknown. Semi-regular radial structure within the ring remains unexplained, and the vertical thickness is apparently comparable with that of the A and B rings. The ring appears to have a preponderance of particles in excess of 10 cm in radius, and the fractional abundance of water ice is much smaller than for the A and B rings. Within the C ring are five known gaps, and two of those gaps, named Colombo and Maxwell, contain eccentric ringlets, possibly associated with undiscovered satellites. The ring mass is estimated to be about 0.000 000 002 times that of Saturn.

Saturn's densest, brightest, and most massive ring is its B ring. It is also the most extensive ring radially other than the tenuous and dusty E ring. While the confining mechanism at the inner boundary is unknown, the outer boundary is caused by a 2:1 resonance with the satellite Mimas, which also gives that boundary a two-lobed shape which turns at the orbital rate of Mimas. There are no clear gaps within the B ring, although the Huygens gap, which contains an eccentric ringlet, is at its outer edge. Water ice is the dominant constituent of the ring, but other material provides coloration. The outer half of the ring is denser and brighter than the inner half for unknown reasons. Before Cassini, there were no complete, high-resolution optical depth profiles for the outer half of the ring. In the middle of this B-ring outer half, ghostly radial spokes occasionally form, possibly as a result of meteoroid impact with ring particles. Material in the spokes becomes electrically charged, and they appear to be electrostatically suspended above the main ring and rotate at the rate of the magnetic field for a time period of from less than one Saturn rotation to nearly three rotations. The outer half of the B ring also contains a myriad of radial structures of an irregular nature, the cause of which is yet to be deciphered. From the sharpness of the ring's outer edge, its thickness has been estimated to be as small as 10 meters! The ring mass is estimated to be 0.000 000 05 times that of Saturn.

Between the B and A rings is the Cassini Division, once thought to be empty, but now known to be filled with material totaling about half the mass of the C ring. It contains six known gaps, the innermost (Huygens gap) and outermost (unnamed) of which are more than 200-km wide and contain ringlets. Its inner boundary is controlled by a 2:1 Mimas resonance, but the source of the sharp increase from the outer Cassini Division to the A ring is not well understood. Like the C ring, the Cassini Division appears to be more polluted with non-icy material than the A and B rings.

The outermost of Cassini's main rings is the A ring. Its outer boundary is at the radial distance of a 7:6 Janus resonance and is seven-lobed. There are two gaps within the A ring, the 325-km wide Encke gap and the 35-km wide Keeler gap. Pan and several partial ringlets (arcs?) occupy the Encke gap. Most of the known density waves and bending waves lie within the A-ring boundaries. Like the B ring, its dominant chemical constituent is water ice. Between the Keeler gap and the outer edge of the A ring, the structure is irregular, both radially and longitudinally. While some of that structure may be related to the gravitational resonance with Janus that is responsible for the seven-lobed outer A-ring edge, the reasons for and detailed processes operative within that irregular structure are not well understood.

Saturn's mysterious F ring is time-variable, both radially and azimuthally. Although the whole width of the ring was seen in stellar occultation data, only the central core was detected by the radio astronomy occultation experiment, implying that there are few particles more than 1–10 cm in radius away from that core, at least at the longitude sampled by that experiment. Bright knots within the F ring may be larger bodies that serve as sources for the F-ring material. At one time, it was thought to be radially confined by the two satellites which flank it, Prometheus and Pandora, but that now seems unlikely. Its mass and vertical thickness are unknown.

The G ring, about 13,000-km wide, does not have well-defined radial boundaries or small-scale radial structure. Its mass is unknown, but its vertical thickness is estimated to be between 100 and 1,000 km. Spacecraft data are not consistent with a narrow range of particle sizes, and the ring has a reddish hue, so the G ring is unlike the E ring in thickness, particle sizes, and composition. Beyond these facts, very little is known about the G ring, including, among other things, its source and its age.

The tenuous and extended E ring has been found to have a bluish color, a very narrow range of particle sizes, and a vertical extent that varies with distance from the planet. Its inner edge is near the orbit of Mimas, but it is both densest and thinnest in vertical extent (about $\pm$ 1,000 km) near the orbit of Enceladus, which is believed to supply the icy material of which it is composed. The volume density of particles diminishes with increasing radial distance beyond Enceladus. The vertical thickness also increases, reaching a vertical half-thickness of at least 15,000 km near the radial distance of Dione. Although they become optically undetectable before it reaches the orbit of Rhea (about 8 $R_S$), E-ring particle concentrations centered near Saturn's equatorial plane may possibly exist nearly to the orbit of Titan (about 20 $R_S$).

Estimates of the age of the Saturn ring system vary from the $4.5 \times 10^9$-year age of Saturn to as little as a few tens of millions of years. It is difficult to understand how a ring system as massive as the satellite Mimas could be so finely divided through natural processes in time periods from tens to hundreds of millions of years, particularly in the absence of an interplanetary meteoroid flux that is thought to have been almost depleted well before that time period. Meteoritic bombardment has other consequences that affect the present composition and structure of the Saturn ring system as well as estimates of its age and evolution. If such bombardment has been ongoing for a large fraction of the age of the solar system, then the composition of the rings should be similar to that of the meteorites, which are not composed of water ice. Even for a moderate ring age, it seems likely that the water-ice particles which largely compose the A and B rings should have been coated with dark meteoritic material, but they remain bright and highly reflective. Meteoritic bombardment should also scatter debris across the ring system, yet there remain what appear to be significant radial variations in the colors of the rings. Other processes also act to deplete the rings of material, either ejecting it from the Saturn system, or, alternatively, causing it to fall into the planet. These and other considerations are discussed in Chapter 10 in light of preliminary findings from the Cassini Orbiter.

## 9.15 CASSINI SCIENTIFIC OBJECTIVES FOR SATURN RING OBSERVATIONS

Cassini ring scientists outlined their primary objectives for Cassini as follows [74]:

(1) Study the configuration of the rings and the dynamic processes responsible for ring structure.
(2) Map the composition and size distribution of ring material.
(3) Investigate the interrelation of rings and moons, including embedded moons.
(4) Determine the distribution of dust and meteoroid distribution in the vicinity of the rings.
(5) Study the interactions between the rings and Saturn's magnetosphere, ionosphere, and atmosphere.

**Table 9.3.** Cassini ring objective contributions by investigation.

| Orbiter investigation | Ring objectives | Web site |
|---|---|---|
| Cassini plasma spectrometer | –, 2, 3, –, – | *http://caps.space.swri.edu/* |
| Cosmic dust analyzer | –, 2, 3, 4, 5 | *http://www.mpi-hd.mpg.de/dustgroup/cassini/* |
| Composite infrared spectrometer | 1, 2, 3, –, 5 | *http://cirs.gsfc.nasa.gov/* |
| Ion/Neutral mass spectrometer | –, 2, –, –, – | *http://caps.space.swri.edu/inms/inms.html* |
| Imaging science | 1, 2, 3, 4, 5 | *http://ciclops.org/index.php* |
| Dual technique magnetometer | 1, –, –, –, – | *http://www3.imperial.ac.uk/spat/research/space missions/cassini/* |
| Magnetosphere imaging instrument | –, –, –, –, 5 | *http://sd-www.jhuapl.edu/CASSINI/* |
| Titan radar | –, –, –, –, 5 | *http://saturn.jpl.nasa.gov/spacecraft/instruments-cassini-radar.cfm* |
| Radio/Plasma wave spectrometer | –, –, –, 4, 5 | *http://www-pw.physics.uiowa.edu/plasma-wave/cassini/home.html* |
| Radio science | 1, 2, 3, –, – | *http://saturn.jpl.nasa.gov/spacecraft/instruments-cassini-rss.cfm* |
| Ultraviolet imaging spectrograph | 1, 2, 3, –, 5 | *http://lasp.colorado.edu/cassini/* |
| Visible/Infrared mapping spectrometer | 1, 2, 3, 4, 5 | *http://wwwvims.lpl.arizona.edu/* |

Without being specific about the observational or analytical methods involved, Table 9.3 outlines which investigations provide a major contribution to the achievement of each of these objectives. As illustrated in the table, each of the 12 Cassini Orbiter investigations contributes to at least one of the ring objectives, and most contribute to several or all of them. Details of the plans and results of each of the 12 investigations can be found on their individual web sites, also shown in Table 9.3.

## 9.16 NOTES AND REFERENCES

[1] Table 9.1 is adapted from table 1 in Cuzzi, J. N., Colwell, J. E., Esposito, L. W., Porco, C. C., Murray, C. D., Nicholson, P. D., Spilker, L. J., Marouf, E. A., French, R. C., Rappaport, N., Muhleman, D., 2003, "Saturn's rings: Pre-Cassini status and mission goals", *Space Science Reviews* **118**, 209–251.

[2] Cuzzi, J. N., Pollack, J. B., 1978, "Saturn's rings: Particle composition and size distribution as constrained by microwave observations", *Icarus* **33**, 233–262.

[3] Figure 9.4 is from Estrada, P. R., Cuzzi, J. N., Showalter, M. R., 2003, "Voyager color photometry of Saturn's main rings: A correction", *Icarus* **166**, 212–222.

[4] Several papers come to the same conclusion about the ring thickness: Focas, J. H., Dollfus, A., 1969, "Propriétés optiques et épaisseur des anneaux de Saturne observés par la tranche

en 1966", *Astronomy Astrophysics* **2**, 251–265; Lumme, K., Irvine, W. M., 1979, "Low tilt angle photometry and the thickness of Saturn's rings", *Astronomy Astrophysics* **71**, 123–130; Brahic, A., Sicardy, B., 1981, "Apparent thickness of Saturn's rings", *Nature* **289**, 447–450; Sicardy, B., Lecacheux, J., Laques, P., Despiau, R., Auge, A., 1982, "Apparent thickness and scattering properties of Saturn's rings from March 1980 observations", *Astronomy Astrophysics* **108**, 296–305.

[5] Burns, J. A., Hamill, P., Cuzzi, J. N., Durisen, R. H., 1979, "On the 'thickness' of Saturn's rings caused by satellite and solar perturbations and by planetary precession", *Astronomical Journal* **84**, 1783–1801.

[6] Shu, F. H., Cuzzi, J. N., Lissauer, J. J., 1983, "Bending waves in Saturn's rings", *Icarus* **53**, 185–206.

[7] Burns, J. A., Hamill, P., Cuzzi, J. N., Durisen, R. H., 1979, "On the 'thickness' of Saturn's rings caused by satellite and solar perturbations and by planetary precession", *Astronomical Journal* **84**, 1783–1801.

[8] Lane, A. L., Hord, C. W., West, R. A., Esposito, L. W., Coffeen, D. L., Sato, M., Simmons, K. E., Pomphrey, R. B., Morris, R. B., 1982, "Photopolarimetry from Voyager 2: Preliminary results on Saturn, Titan, and the rings", *Science* **215**, 537–543.

[9] A coherent radio signal is one that is composed of a tightly controlled frequency (from an ultra-stable oscillator). Generally, there are also other controlled factors, such as the phase and polarization of the signal. A coherent radio signal is the radio equivalent of a laser beam in optics.

[10] Diffraction is a phenomenon that occurs when a coherent radio or optical beam passes a sharp opaque edge, like a knife edge. The radiation is "scattered" in a way that creates (in the case of optical diffraction) a pattern of alternating dark and light bands near the edge of the shadow.

[11] Marouf, E. A., Tyler, G. L., 1982, "Microwave edge diffraction in Saturn's ring observations with Voyager 1", *Science* **217**, 243–245; Tyler, G. L., Marouf, E. A., Simpson, R. A., Zebker, H. A., Eshleman, V. R., 1983, "The microwave opacity of Saturn's rings at wavelengths of 3.6 and 13 cm from Voyager 1 radio occultation", *Icarus* **54**, 160–188.

[12] Borderies, R., Goldreich, P., Tremaine, S., 1982, "Sharp edges of planetary rings", *Nature* **299**, 209–211.

[13] Bobrov, M. S., 1970, "The Rings of Saturn" (translation), NASA TTF-701; Bobrov, M. S., "Physical properties of Saturn's rings", in *Surfaces and Interiors of Planets and Satellites*, edited by Dollfus, pp. 377–458.

[14] Cuzzi, J. N., Durisen, R. H., Burns, J. A., Hamill, P., 1979, "The vertical structure and thickness of Saturn's rings", *Icarus* **38**, 54–68; Cuzzi, J. N., Burns, J. A., Durisen, R. H., Hamill, P. M., 1979, "The vertical structure and thickness of Saturn's rings", *Nature* **281**, 202–204; Brahic, A., Sicardy, B., 1981, "Apparent thickness of Saturn's rings", *Nature* **289**, 447–450; Stewart, G. R., Lin, D. N. C., Bodenheimer, P., 1984, "Collision-induced transport processes in planetary rings", in *Planetary Rings*, edited by Greenberg and Brahic, pp. 447–512.

[15] Hapke, B., 1963, "A theoretical photometric function for the lunar surface", *Journal of Geophysical Research* **68**, 4571–4586; Hapke, B., 1981, "Bidirectional reflectance spectroscopy. I. Theory", *Journal of Geophysical Research* **86**, 3039–3054; Irvine, W. M., 1966, "The shadowing effect in diffuse reflection", *Journal of Geophysical Research* **71**, 2931–2937.

[16] Marouf, E. A., Tyler, G. L., Zebker, H. A., Eshleman, V. R., 1983, "Particle size distribution in Saturn's rings from Voyager 1 radio occultation", *Icarus* **49**, 161–193;

Zebker, H. A., Marouf, E. A., Tyler, G. L., 1985, "Saturn's rings: Particle size distributions for thin layer models", *Icarus* **64**, 531–548.

[17] Ostro, S. J., Pettengill, G. H., Campbell, D. B., 1980, "Radar observations of Saturn's rings at intermediate tilt angles", *Icarus* **41**, 381–388; Ostro, S., Pettengill, G., 1983, "A review of radar observations of Saturn's rings," Proceedings of *I.A.U. Colloquium 75 Planetary Rings*, edited by Brahic.

[18] Baum, W. A., Kreidl, T., Westphal, J., Danielson, G. E., Seidelmann, P. K., Pascu, D., Currie, D. G., 1981, "Saturn's E Ring. I: CCD observations of March 1980". *Icarus* **47**, 84–96; Lamy, P., Mauron, N., 1981, "Observations of Saturn's outer ring and new satellites during the 1980 edge-on presentation", *Icarus* **46**, 181–186; De Pater, I., Martin, S. C., Showalter, M. R., 2004, "Keck near-infrared observations of Saturn's E and G rings during Earth's ring plane crossing in August 1995", *Icarus* **172**, 446–454.

[19] Nicholson, P. D., Showalter, M. R., Luke Dones, L, French, R. G., Larson, S. M., Lissauer, J. J., McGhee, C. A., Seitzer, P., Sicardy, B. G., Edward Danielson, G. E., "Observations of Saturn's ring-plane crossings in August and November 1995", *Science* **272**, 509–515.

[20] Aubier, M. G., Meyer-Vernet, N., Pedersen, B. M., 1983, "Shot noise from grain and particle impacts in Saturn's ring plane", *Geophysical Research Letters* **10**, 5–8; Gurnett, D. A., Grün, E., Gallagher, D., Kurth, W. S., Scarf, F. L., 1983, "Micron-sized particles detected near Saturn by the Voyager plasma-wave experiment", *Icarus* **53**, 236–254.

[21] Camichel, H., 1958, "Mesures photométriques de Saturne et de son anneau", *Annals of Astrophysics* **21**, 231–242.

[22] Ferrin, I., 1975. "On the structure of Saturn's rings and the 'real' rotational period for the planet", *Astrophysics and Space Science* **33**, 453–457; Reitsema, H. J., Beebe, R. F., Smith, B. A., 1976, "Azimuthal brightness variations in Saturn's rings", *Astronomical Journal* **81**, 209–215.

[23] Lumme, K., Irvine, W. M., 1979, "A model for the azimuthal brightness variations in Saturn's rings", *Nature* **282**, 695–696; Lumme, K., Irvine, W. M., 1979, "Azimuthal brightness variations of Saturn's rings. III. Observations at tilt angle B approximately equal to 11.5°", *Astrophysical Journal* **229**, L109–L111; Lumme, K., Esposito, L. W., Irvine, W. M., Baum, W. A., 1977, "Azimuthal brightness variations of Saturn's rings. II. Observations at intermediate tilt angle", *Astrophysical Journal* **216**, L123–L126; Thompson, W. T., Lumme, K., Irvine, W. M., Baum, W. A., Esposito, L. W., 1981, "Saturn's rings—azimuthal variations, phase curves, and radial profiles in four colors", *Icarus* **46**, 187–200.

[24] Dones, L., Cuzzi, J. N., Showalter, M. R., 1993, "Voyager photometry of Saturn's A ring", *Icarus* **105**, 184–215.

[25] Marouf, E. A., Tyler, G. L., Zebker, H. A., Simpson, R. A., Eshleman, V. R., 1983, "Particle size distributions in Saturn's rings from Voyager 1 radio occultation", *Icarus* **54**, 189–211.

[26] Salo, H., Karjalainen, R., French, R. G., 2004, "Photometric modeling of Saturn's rings. II. Azimuthal asymmetry in reflected and transmitted light", *Icarus* **170**, 70–90. The suggestion that the A-ring brightness asymmetry might be due to linear strings of particles related to density waves was first made by Colombo, G., Goldreich, P., Harris, A. W., 1976, "Spiral structure as an explanation for the asymmetric brightness of Saturn's A ring", *Nature* **264**, 344–345.

[27] Showalter, M. R., 1998, "Detection of impact events in Saturn's F ring", *Science* **282**, 1099–1102.

[28] French, R. G., McGhee, C. A., Dones, L., Lissauer, J., 2002, "Saturn's wayward shepherds: The peregrinations of Pandora and Prometheus", *Icarus* **162**, 143–170.

[29] McGhee, C. A., French, R. G., Dones, L., Cuzzi, J. N., Salo, H. J., Danos, R., 2005, "HST observations of spokes in Saturn's B ring", *Icarus* **173**, 508–521.

[30] Prometheus and Pandora, which flank the F ring, undergo sporadic and almost instantaneous changes to their orbit parameters. The precise cause is unknown. This behavior is discussed in the following three articles: French *et al.*, 2003 [27]; Murray, C. D., Giuliatti-Winter, S. M., 1996, "Periodic collisions between the moon Prometheus and Saturn's F ring", *Icarus* **129**, 304–316; Goldreich, P., Rappaport, N., 2003, "Chaotic motions of Prometheus and Pandora", *Icarus* **162**, 391–399. Janus and Epimetheus very nearly share the same orbit. However, the slight difference in their orbital periods means that the inner moon will eventually lap the outer moon. The gravitational interaction that occurs when they approach each other slows the inner moon, causing it to move outward, and speeds up the outer moon, causing it to move inward. A recent such exchange was observed by the Cassini Orbiter in January 2006, when Janus's orbital radius decreased by about 20 km and Epimetheus's orbital radius increased by about 80 km, thereby exchanging their designations as the inner or outer member of the pair. The orbit of Janus is affected less than that of Epimetheus because Janus is approximately four times as massive as Epimetheus.

[31] The existence of Pan was predicted on the basis of observed wakes by Cuzzi, J. N., Scargle, J. D., 1985, "Wavy edges suggest moonlet in Encke's gap", *Astrophysical Journal* **292**, 276–290. The actual discovery of Pan in Voyager images was reported by Showalter, M. R., 1991, "Visual detection of 1981 S13, Saturn's 18th satellite, and its role in the Encke gap", *Nature* **351**, 709–713.

[32] Cuzzi, J. N., Lissauer, J. J., Esposito, L. W., Holberg, J. B., Marouf, E. A., Tyler, G. L., Boischot, A., 1984, "Saturn's rings: Properties and processes", in *Planetary Rings*, edited by Greenberg and Brahic, pp. 73–199.

[33] Broadfoot, A. L., Sandel, B. R., Shemansky, D. E., Holberg, J. B., Smith, G. R., Strobel, D. F., McConnell, J. C., Kumar, S., Hunten, D. M., Atreya, S. K., Donahue, T. M., Moos, H. W., Bertaux, J. L., Blamont, J. E., Pomphrey, R. B., Linick, S., 1981, "Extreme ultraviolet observations from Voyager 1 encounter with Saturn", *Science* **212**, 206–211.

[34] Morfill, G. E., Fechtig, H., Grün, E., Goertz, C. E., 1983, "Some consequences of meteoroid impacts on Saturn's rings", *Icarus* **55**, 439–447.

[35] Shemansky, D. E., Smith, G. R., 1982, "Whence comes the 'Titan' hydrogen torus?", *Eos* **63**, 1019; Shemansky, D. E., Ajello, J. M., 1983, "The Saturn spectrum in the EUV: Electron-excited hydrogen", *Journal of Geophysical Research* **88**, 459–464.

[36] Goldstein, R. M., Morris, G. A., 1973, "Radar observations of the rings of Saturn", *Icarus* **20**, 260–262; Pollack, J. B, Summers, A., Baldwin, B., 1973, "Estimates of the size of the particles in the rings of Saturn and their cosmogonic implications", *Icarus* **20**, 263–278.

[37] Cuzzi, J. N., Pollack, J. B., 1978, "Saturn's rings: Particle composition and size distribution as constrained by microwave observations. I: Radar observations", *Icarus* **33**, 233–263; Ostro, S. J., Pettengill, G. B., Campbell, D. B., 1980, "Radar observations of Saturn's rings at intermediate tilt angles", *Icarus* **41**, 381–388.

[38] Tyler, G. L., Marouf, E. A., Simpson, H. A., Eshleman, V. R., 1983, "The microwave opacity of Saturn's rings at wavelengths of 3.6 and 13 cm from Voyager 1 radio occultation", *Icarus* **54**, 160–188; Marouf, E. A., Tyler, G. L., Zebker, H. A., Simpson, R. A., Eshleman, V. R., 1983, "Particle size distribution in Saturn's rings from Voyager 1 radio occultation", *Icarus* **54**, 189–211; Zebker, H. A., Marouf, E. A., Tyler, G. L., 1985, "Saturn's ring particle size distribution for thin layer models", *Icarus* **64**, 531–548.

[39] French, R. G., Nicholson, P. D., 2000, "Saturn's rings II: Particle sizes inferred from stellar occultation data", *Icarus* **145**, 502–523.

[40] Epstein, E. E., Janssen, M. A., Cuzzi, J. N., 1984, "Saturn's rings—3-mm low-inclination observations and derived properties", *Icarus* **58**, 403–411; Grossman, A., 1991, "Microwave imaging of Saturn's deep atmosphere and rings", PhD dissertation, California Institute of Technology.

[41] Cuzzi, J. N., Burns, J. A., 1988, "Charged particle depletion surrounding Saturn's F ring: Evidence for a moonlet belt", *Icarus* **74**, 284–324.

[42] Showalter, M. R., 1998, "Detection of impact events in Saturn's F ring", *Science* **282**, 1099–1102.

[43] *Ibid.*; Bosh, A., Rivkin, A. S., 1996, "Observations of Saturn's inner satellites during the May 1995 ring-plane crossing", *Science* **272**, 518–521; Nicholson, P. D., Showalter, M. R., Dones, L., French, R. G., Larson, S. M., Lissauer, J. J., McGhee, C. A., Seitzer, P., Sicardy, B., Danielson, G. E., 1996, "Observations of Saturn's ring-plane crossings in August and November 1995", *Science* **272**, 509–515.

[44] Showalter, M. R., Cuzzi, J. N., 1993, "Seeing ghosts: Photometry of Saturn's G ring", *Icarus* **103**, 124–143; van Allen, J. A., 1982, "Findings on rings and inner satellites of Saturn by Pioneer 11"; *Icarus* **51**, 509–527.

[45] Nicholson, P. D., Showalter, M. R., Dones, L., French, R. G., Larson, S. M., Lissauer, J. J., McGhee, C. A., Seitzer, P., Sicardy, B., Danielson, G. E., 1996, "Observations of Saturn's ring-plane crossings in August and November 1995", *Science* **272**, 509–515.

[46] Throop, H. B., Esposito, L. W., 1998, "G ring particle sizes derived from ring plane crossing observations", *Icarus* **131**, 152–166.

[47] Meyer-Vernet, N., Lecacheux, A., Pedersen, B. M., 1998, "Constraints on Saturn's G ring from the Voyager 2 radio astronomy experiment", *Icarus* **132**, 311–320.

[48] See, for example, fig. 7 of Cuzzi, J. N., Lissauer, J. J., Esposito, L. W., Holberg, J. B., Marouf, E. A., Tyler, G. L., Boischot, A., 1984, "Saturn's rings: Properties and processes", in *Planetary Rings*, edited by Greenberg and Brahic, p. 101.

[49] Esposito, L. W., Cuzzi, J. N., Holberg, J. H., Marouf, E. A., Tyler, G. L., Porco, C. C., 1984, "Saturn's rings: Structure, dynamics, and particle properties", in *Saturn*, edited by Gehrels and Matthews, pp. 463–545 (and references therein).

[50] Ingersoll, A. P., Neugebauer, G., Orton, G. S., Münch, G., Chase, S. C., 1980, "Pioneer Saturn infrared radiometer—preliminary results", *Science* **207**, 439–443; Tokunaga, A. T., Caldwell, J., Nolt, I. G., 1980, "The 20-micron brightness temperature of the unilluminated side of Saturn's rings", *Nature* **287**, 212–214; Froidevaux, L., Ingersoll, A. P., 1980, "Temperatures and optical depths of Saturn's rings and a brightness temperature for Titan", *Journal of Geophysical Research* **85**, 5929–5936.

[51] Voyager eclipse cooling data from Voyager 1 is discussed by Hanel, R., Conrath, B., Flasar, F. M., Kunde, V., Maguire, W., Pearl, J., Pirraglia, J., Samuelson, R., Herath, L., Allison, M., Cruikshank, D., Gautier, D., Gierasch, P., Horn, L., Koppany, R., Ponnamperuma, C., 1981, "Infrared observations of the Saturnian system from Voyager 1", *Science* **212**, 192–200. Earlier observations from Earth are discussed by Petit, E., 1961, "Planetary temperature measurements", in *The Solar System*, Vol. III, edited by Kuiper and Middlehurst, pp. 400–428.

[52] French, R. G., Nicholson, P. D., Cooke, M. J., Elliot, J. L., Matthews, K., Perkovic, O., Tollestrup, E., Harvey, P., Chanover, N. J., Clark, M. A., Dunham, E. W., Forrest, W., Harrington, J., Pipher, J., Brahic, A., Grenier, I., Roques, F., Arndt, M., 1993, "Geometry of the Saturn system from the 3 July 1989 occultation of 28 Sgr and Voyager observations", *Icarus* **103**, 163–214; Harrington, J., Cooke, M. L., Forrest, W. J., Pipher, J. L., Dunham,

E. W., Elliot, J. L., 1993, "IRTF observations of the occultation of 28 Sgr by Saturn", *Icarus* **103**, 235–252; Hubbard, W. B., Porco, C. C., Hunten, D. M., Rieke, G. H., McCarthy, D. W., Haemmerle, V., Clark, R., Turtle, E. P., Haller, J., McLeod, B., Lebofsky, L. A., Marcialis, R., Holberg, J. B., Landau, R., Carrasco, L., Elias, J., Buie, M. W., Persson, S. E., Boroson, T., West, S., Mink, D. J., 1993, "The occultation of 28 Sgr by Saturn—Saturn pole position and astrometry", *Icarus* **103**, 215–234.

[53] Results from one such occultation are reported by Elliot, J. L., Bosh, A. S., Cooke, M. L., Bless, R. C., Nelson, M. J., Percival, J. W., Taylor, M. J., Dolan, J. F., Robinson, E. L., van Critters, G. W., 1993, "An occultation by Saturn's rings on 1991 October 2–3 observed with HST", *Astronomical Journal* **106**, 2544–2572.

[54] Showalter, M. R., 1998, "Detection of impact events in Saturn's F ring", *Science* **282**, 1099–1102.

[55] Throop, H. B., Esposito, L. W., 1998, "G ring particle sizes derived from ring plane crossing observations", *Icarus* **131**, 152–166.

[56] Bosh, A., Rivkin, A. S., 1996, "Observations of Saturn's inner satellites during the May 1995 ring-plane crossing", *Science* **272**, 518–521; Nicholson, P. D., Showalter, M. R., Dones, L., French, R. G., Larson, S. M., Lissauer, J. J., McGhee, C. A., Seitzer, P., Sicardy, B., Danielson, G. E., 1996, "Observations of Saturn's ring-plane crossings in August and November 1995", *Science* **272**, 509–515; French, R. G., McGhee, C. A., Dones, L., Lissauer, J., 2003, "Saturn's wayward shepherds: Pandora and Prometheus", *Icarus* **162**, 143–170.

[57] The Doppler effect was named for Christian J. Doppler, who showed that motion of a light-, radio-, or sound-wave source toward (or away from) an observer will cause the observed waves to be shifted to higher (or lower) frequencies. He also showed that the magnitude of the shift was directly proportional to the speed of the source relative to the observer. The Doppler effect is responsible for the apparent reduction in pitch in the whistle of a rapidly moving train as it passes by a stationary listener standing at a railroad crossing. It is also by means of the red-shift (Doppler shift toward longer wavelengths) of light received from distant galaxies that astronomers know that our universe is expanding.

[58] Nicholson, P. D., French, R. G., Campbell, D. B., Margot, J.-L., Nolan, M. C., Black, G. J., Salo, H., 2005, "Radar imaging of Saturn's rings", *Icarus* **177**, 32–62.

[59] French, R. G., Cuzzi, J., Danos, R., Dones, L., Lissauer, J., 1998, "Hubble space telescope observations of spokes in Saturn's rings", proceedings of *The Jovian system after Galileo, the Saturnian system before Cassini-Huygens*, Nantes, France, 11–15 May 1998; Cuzzi, J. N., French, R. C., Dones, L., 2002, "HST multicolor photometry (255–1042 nm) of Saturn's main rings. I: Radial profiles, phase and opening angle variations, and regional spectra", *Icarus* **158**, 199–223.

[60] Poulet, F., Cruikshank, D. P., Cuzzi, J. N., Roush, T., French, R. C., 2003, "Compositions of Saturn's rings A, B, and C from high resolution near-infrared spectroscopic observations", *Astronomy and Astrophysics* **412**, 305–316.

[61] See, for example, Esposito, L. W., Cuzzi, J. N., Holberg, J. H., Marouf, E. A., Tyler, G. L., Porco, C. C., 1984, "Saturn's rings: Structure, dynamics, and particle properties", in *Saturn*, edited by Gehrels and Matthews, pp. 463–545; Cuzzi, J. N., Lissauer, J. J., Esposito, L. W., Holberg, J. B., Marouf, E. A., Tyler, G. L., Boischot, A., 1984, "Saturn's rings: Properties and processes", in *Planetary Rings*, edited by Greenberg and Brahic, pp. 73–199.

[62] Eshleman, V. R., Tyler, G. L., Anderson, J. D., Fjeldbo, G., Levy, G. S., Wood, G. E., Croft, T. A., 1977, "Radio science investigations with Voyager", *Space Science Reviews* **21**, 207–232.

[63] Marouf, E. A., Tyler, G. L., Eshleman, V. R., 1982, "Theory of radio occultation by Saturn's rings", *Icarus* **49**, 161–193.
[64] The amount of radiation received at a radio or optical detector after transmission through Saturn's rings (or other similar particulate target) is a combination of directly transmitted light and diffracted (scattered) light from adjacent regions. For non-coherent light (like starlight being viewed through a ring) the two effects are essentially indistinguishable, and the ring appears less opaque than it actually is. For coherent radiation (like Voyager's or Cassini's radio beams or like an optical laser), the diffracted light is shifted in frequency or phase relative to the directly transmitted light and thus can be distinguished and removed. For targets with relatively low optical (or radio) depth, the diffracted or scattered signal is equal in strength to the direct signal (for both coherent and non-coherent radiation), so the true optical (or radio) depth calculated for the coherent radiation (with diffraction effects removed) is twice that of the non-coherent radiation, a fact that must be kept in mind when comparing optical depths derived from non-coherent light to radio depths derived from coherent radio signals. For optically thick rings, multiple scattering or diffraction (in which a single ray of radiation "bounces" off more than one particle on its way to the detector) may become more important, and the ratio between derived coherent radio and non-coherent optical depths deviates from a two-to-one ratio by some unknown amount.
[65] Marouf, E. A., Tyler, G. L., Zebker, H. A., Simpson, R. A., Eshleman, V. R., 1983, "Particle size distributions in Saturn's rings from Voyager 1 radio occultation", *Icarus* **54**, 189–211.
[66] Tyler, G. L., Marouf, E. A., Simpson, R. A., Zebker, H. A., Eshleman, V. R., 1983, "The microwave opacity of Saturn's rings at wavelengths of 3.6 and 13 cm from Voyager 1 radio occultation", *Icarus* **54**, 160–188; Marouf, E. A., Tyler, G. L., Rosen, P. A., 1986, "Profiling Saturn's rings by radio occultation", *Icarus* **68**, 120–166.
[67] Marouf, E. A., Tyler, G. L., Zebker, H. A., Simpson, R. A., Eshleman, V. R., 1983, "Particle size distributions in Saturn's rings from Voyager 1 radio occultation", *Icarus* **54**, 189–211.
[68] Van Allen, J. A., 1982, "Findings on rings and inner satellites of Saturn by Pioneer 11", *Icarus* **51**, 509–527; Van Allen, J. A., 1983, "Absorption of energetic protons by Saturn's G ring", *Journal of Geophysical Research* **88**, 6911–6918; Van Allen, J. A., 1984, "Energetic particles in the inner magnetosphere of Saturn", in *Saturn*, edited by Gehrels and Matthews, pp. 281–317.
[69] Gehrels, T., Baker, L. R., Beshore, E., Blenman, C., Burke, J. J., Castillo, N. D., Da Costa, B., Degewij, J., Doose, L. R., Fountain, J. W., Gotobed, J., KenKnight, C. E., Kingston, R., McLaughlin, G., McMillan, R., Murphy, R., Smith, P. H., Stoll, C. P., Strickland, R. N., Tomasko, M. G., Wijesinghe, M. P., Coffeen, D. L., Esposito, L. W., 1980, "Imaging Photopolarimeter on Pioneer Saturn", *Science* **207**, 434–439.
[70] Chenette, D. L., Cooper, J. F., Eraker, J. H., Pyle, K. E., Simpson, J. A., 1980, "High-energy trapped radiation penetrating the rings of Saturn", *Journal of Geophysical Research* **85**, 5785–5792; Fillius, W., McIlwain, C., 1980, "Very energetic protons in Saturn's radiation belt", *Journal of Geophysical Research* **85**, 5803–5811.
[71] Goertz, C. K., Morfill, G. E., 1983, "A model for the formation of spokes in Saturn's rings", *Icarus* **53**, 219–229; Goertz, C., "Formation of Saturn's spokes", *Advances in Space Research* **4**(9), 137–141.
[72] Cuzzi, J. N., Durisen, R. H., 1990, "Meteoroid bombardment of planetary rings. General formulation and effects of Oort cloud projectiles", *Icarus* **84**, 467–501.
[73] The ring objectives for the Cassini mission are outlined on the Cassini home page: *http://saturn.jpl.nasa.gov/science/rings.cfm*

[74] Ring objectives for the Cassini Orbiter mission are documented in several sources, but perhaps the most easily accessible is the Cassini-Huygens home page: *http://saturn.jpl.nasa.gov/science/rings.cfm.*

## 9.17 BIBLIOGRAPHY

Cuzzi, J. N., Colwell, J. E., Esposito, L. W., Porco, C. C., Murray, C. D., Nicholson, P. D., Spilker, L. J., Marouf, E. A., French, R. C., Rappaport, N., Muhleman, D., 2002, "Saturn's rings: Pre-Cassini status and mission goals", *Space Science Reviews* **118**, 209–251.

*Planetary Rings*, 1984, University of Arizona Press, edited by Richard Greenberg and André Brahic. Note especially the following chapters: "Saturn's rings: Properties and processes" (pp. 73–199), "The ethereal rings of Jupiter and Saturn" (pp. 200–272), "Dust–magnetosphere interactions" (pp. 275–332), "Effects of radiation forces on dust particles" (pp. 333–366), "Ring particles: Collisional interactions and physical nature" (pp. 367–415), "Waves in planetary rings" (pp. 513–561), "Ring particle dynamics in resonances" (pp. 562–588), and "Unsolved problems in planetary ring dynamics" (pp. 713–734).

*Saturn*, 1984, University of Arizona Press, edited by Tom Gehrels and Mildred Shapley Matthews. Note especially the following chapters: "Saturn's ring: Structure, dynamics, and particle properties" (pp. 463–545), and "Electrodynamic processes in the ring system of Saturn" (pp. 546–589).

## 9.18 PICTURES AND DIAGRAMS

Figure 9.1 *http://antwrp.gsfc.nasa.gov/apod/image/f_ring_vg1.gif*
Figure 9.2 *http://photojournal.jpl.nasa.gov/catalog/PIA02283*
Figure 9.3 From Estrada *et al.* [3]; reproduced in Cuzzi *et al.* [1].
Figure 9.4 *http://www2.jpl.nasa.gov/saturn/gif/saturn_plot.gif*
Figure 9.5 *http://photojournal.jpl.nasa.gov/jpeg/PIA04913.jpg*
Figure 9.6 *http://photojournal.jpl.nasa.gov/jpeg/PIA00335.jpg*
Figure 9.7 From Salo *et al.* [26] (fig. 3).
Figure 9.8 *http://photojournal.jpl.nasa.gov/jpeg/PIA02274.jpg*
Figure 9.9 Voyager 2 image 260-1473.
Figure 9.10 From Cuzzi and Scargle [31] (fig. 1).
Figure 9.11 *http://photojournal.jpl.nasa.gov/jpeg/PIA01388.jpg*
Figure 9.12 *http://photojournal.jpl.nasa.gov/jpeg/PIA01964.jpg*
Figure 9.13 *http://www.naic.edu/~pradar/radarpage.html*
Figure 9.14 *http://photojournal.jpl.nasa.gov/jpeg/PIA02275.jpg*

# 10

# Early results about Saturn's rings from Cassini

## 10.1 OVERVIEW

The Voyager 1 and 2 encounters opened the eyes of the world to the fascinating complexity of Saturn's rings. For the more than two decades since, the Voyager images have remained fixed in our minds as if frozen in a lightning flash, and it hasn't been until recently that Cassini has shown that these observations represent snapshots of a very changeable, dynamic system—with many features in the rings varying on timescales of years, months, and even days. Cassini has made a number of other new discoveries as well, concerning the ring-particle composition and size distribution, and how they vary from place to place. As this chapter goes to press with less than half of Cassini's planned ring observations being completed, many surprises are yet to come and many puzzles will be resolved.

The interaction between the rings and nearby moons includes the physics of spiral waves driven at orbit resonances with moons (see Section 9.5). Cassini has garnered many new observations of these waves (Section 10.3). Related to the interaction between distant moons and rings that causes spiral waves is the case where a smaller, but closer, moonlet affects ring material. The moonlet may even be embedded within the rings (Section 9.5), in which case it can clear an empty gap. Detailed analysis during the decade after the Voyager encounters did lead to the discovery of one moonlet in the Encke gap, and Cassini observations have now revealed another (Section 10.5). The same physics—by which the gravity of a local moonlet actually pushes ring material away from it—was once thought to explain how narrow ringlets were confined or prevented from spreading out. The term "shepherd moons" was applied to moons that straddled and confined a ringlet, such as the narrow, stranded and kinky F ring lying just outside Saturn's main rings, because they were thought to act like sheepdogs confining an unruly flock of sheep [1]. However, over the years we have realized that the so-called shepherds' mutual interactions make their orbits "chaotic", undergoing occasional glitches and jumps [2]. It is likely that the entire region between these moonlets, including the F ring itself and any little shards and small objects that populate the region, is also permeated by orbital chaos rather than stable confinement. Do these chaotic orbits lead to occasional collisions and new ringlets [3]? Some tantalizing hints have been provided by analysis of old Voyager data, showing formation of new clumps on weekly or monthly timescales [4], and from Hubble Space Telescope observations of new clumps, observed fleetingly when Saturn's rings turned edge-on to the Earth in 1995 [5]. Some new Cassini observations are already revealing dramatic time evolution in the entire F-ring region (Section 10.5).

Most of the known structure in the rings remains a puzzle; for example, structure having no clear pattern or preferred length-scale fills Saturn's B ring and inner A ring—the most optically thick parts of the rings—and is known as *irregular structure*. It is only found in the optically thicker parts of the rings; current theories may be able to explain some of it, but certainly not all or even most of it. How opaque is it? What kinds of structures or perhaps buried moonlets lurk in its depths? How does it vary with angle around the rings? Cassini has now conducted nine optimized radio occultations that penetrate nearly all of the densest parts of the rings, revealing new kinds of structure. Moreover, Cassini has observed several stellar occultations that also penetrate this structure along various lines of sight. These results will be discussed in Section 10.3.

A different kind of structure is too minuscule in scale to be directly observed even with Cassini's instruments, but manifests itself indirectly in several ways which are now under study. Section 9.4 describes how the brightness of the A ring varies with orbit longitude; this appearance is thought to result from transient streamers or "gravity wakes" [6], formed when clumps of particles began to collapse under their own self-gravity but were just as quickly sheared out by their differential rotation [7]. These shearing clumps have about the same scale as the ring thickness—tens of meters. This dynamical clumping into temporary "superparticles" creates small-scale local variances that confuse the issue of what the "typical" packing density, or the size distribution, of the actual ring particles are. However, Cassini observations are now starting to constrain the properties of the wake effect, the vertical thickness and packing density, and the underlying ring-particle size distribution, in detail, as we discuss in Section 10.3.

This gravity-wake effect is connected to the fundamental dynamical question of the collisional or random velocity of the ring particles, which is at the heart of all the dynamics that causes their structure. The collisional velocity is manifested in the ring vertical thickness (Section 9.3) and packing density of the particles, which can, in principle, be constrained in several ways by remote observations. Interparticle collisions give the rings a viscous nature (Section 9.5), which enters into all the physics of spiral waves, shepherding, ring spreading, etc. Indirect determinations of the particle random velocities come from a variety of different Cassini observations (Section 10.3).

Besides deciphering the *structure* of the rings, we also need to understand the *composition* of the underlying ring particles. As described in Section 9.8, water ice is the main constituent of the rings [8]. However, the general tawny color of the rings tells us that other materials, in addition to ice, must be widespread, even if only in trace amounts. Also, the brightness (albedo, or reflectivity) of the ring particles, as well as their color, does vary from place to place in the rings, indicating that their composition also varies with location [9]. Cassini carries several new capabilities for measuring composition remotely, and its extended tour presents the variety of geometrical opportunities needed to separate out compositional variations from variations in, for instance, abundance or optical depth (Section 10.4).

How ring-particle composition varies from region to region can tell us whether the rings formed of material with a uniform composition, or in several possibly overlapping bands of different composition, but we need to understand first if there has been blurring and spreading of material from one place to another. Particles can swap material in collisions, and so parcels of matter with different composition can slowly diffuse through the rings. More important, probably, is spreading around of material ejected by high-speed impacts onto the rings by interplanetary meteoroids [10] (Section 10.6). Models suggest this is to be expected and is potentially a diagnostic of the ring age.

These models of the process of meteoroid bombardment find it to be potentially of great importance, but dependent on several poorly known parameters. The importance comes from the fact that the rings have such an enormous area for their relatively puny mass—the area/mass ratio for the rings is about a million times larger than for, say, Earth. We don't really know what the meteoroid mass flux is at Saturn, unfortunately, but estimates have been made which indicate that the rings might have swept up their own mass in meteoroids over the age of the solar system! This would obviously change both their dynamics and their composition dramatically. No icy particle could absorb an equal mass of dark, carbon-rich meteoroid material without becoming nearly as dark as charcoal (such as the Uranus and Neptune rings—Section 11.4). Thus, the fresh, icy surfaces of most of Saturn's ring particles might be telling us that Saturn's rings aren't as old as the solar system after all. In Section 10.6 we review other ways in which Cassini observations will help us understand this process better.

A young age for the rings, implied by both their dynamical interactions with nearby moons (Section 9.5) and their evolution under meteoroid bombardment, remains controversial because of many uncertainties in both lines of argument. The arguments are independent, and they both give about the same result, but perhaps this is a coincidence. There are difficulties with envisioning just how such a massive ring system could have formed so recently, after all the heavy bombardment of the very early solar system had long since died down. Cassini will help to resolve this issue in several ways, helping to bring the age and origin of Saturn's rings into sharper focus (Sections 10.6 and 10.7).

## 10.2 CASSINI SPACECRAFT AND MISSION

The Cassini Orbiter carried and deposited ESA's Huygens Probe into Titan's atmosphere on January 14, 2006, and is the home for all the instruments doing ring observations. The Orbiter is 11 m long and its high-gain telemetry antenna is 4 m across (Figure 10.1). All of its science instruments are fastened to the spacecraft,

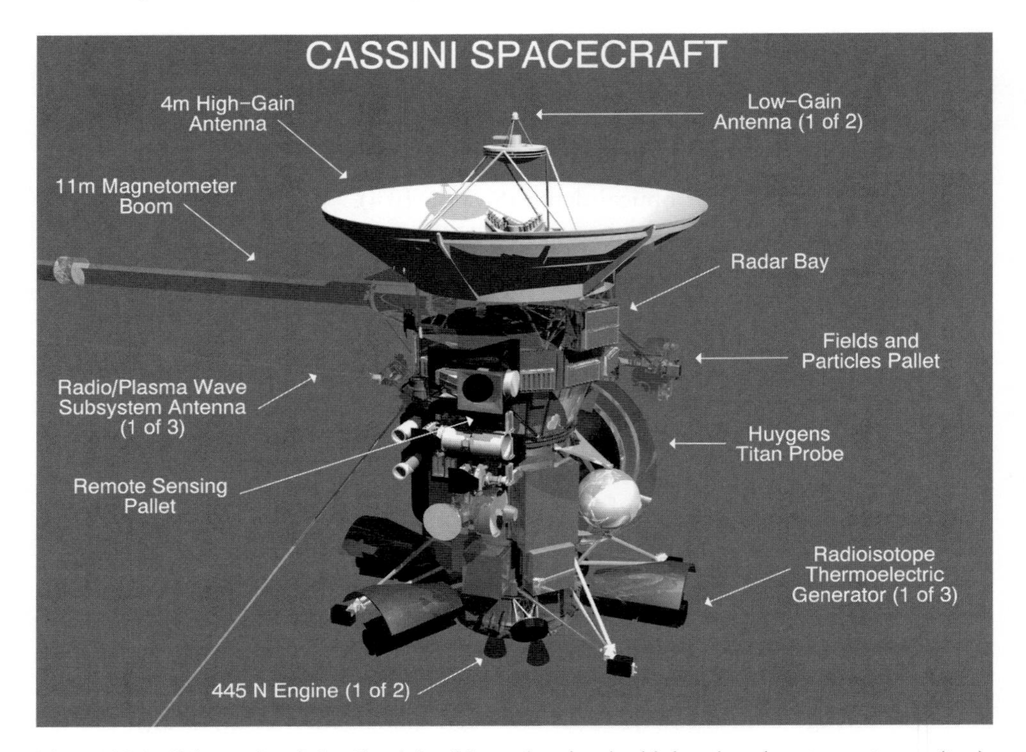


**Figure 10.1.** Schematic of the Cassini orbiter, showing its high-gain telemetry antenna (top), the remote-sensing pallet which includes the ISS, VIMS, CIRS, and UVIS telescopes (front center), and other instruments. The Huygens entry probe is shown at the rear. Also of interest for some ring observations is the RPWS antenna.

although some have internal scanning capability. The main investigations for ring observations are the four optical remote sensing (ORS) instruments fastened to the side of the spacecraft, and the high-gain antenna itself. The ORS instruments include the narrow-angle (high-resolution) and wide-angle cameras of the imaging science subsystem (ISS), which can take pictures in up to 16 different sets of color filters; the visible and near-infrared mapping spectrometer, or VIMS, which scans portions of the rings to return relatively coarse images in hundreds of spectral bands at once; the thermal infrared instrument (CIRS), which measures the temperatures of the ring particles as they move from day to night and the lit face of the rings to the unlit face; and an ultraviolet imaging spectrograph (UVIS). UVIS and VIMS are also designed to conduct dozens of stellar *occultations* by the rings (and planet), in which the brightness of a star is monitored every few milliseconds as it is covered up, or occulted, by the rings or some other target. Because of the tiny effective size of the star, we measure the structure of the rings it shines through on ultra-fine scales (tens of meters) in this manner. Cassini's four-meter radio antenna not only conducts similar occultations (except that the radio source is on the spacecraft and the receiving equipment is on Earth) at *radio* wavelengths, very powerful for measuring how the size distribution varies across all this fine-scale structure, but it can act as a normal radio telescope as well.

Other instruments also contribute to ring science goals in important ways. The Cosmic Dust Analyzer (CDA) detects the mass and composition of dust grains in the micron-size range which happen to hit it, especially important for the diffuse E and G rings where many of the particles are detectable primarily this way. Moreover, it is hoped that CDA can contribute to determining the mass flux of meteoroids into the Saturn system. While able to measure the velocity, direction, and composition of particles that hit it, the CDA has a very small collecting area, so it tends not to detect many particles. Another instrument makes use of the much larger area of the entire spacecraft as a collector, by measuring a voltage pulse from each puff of charged gas that results when a tiny grain hits the spacecraft at high speed [11]—the radio and plasma wave spectrometer (RPWS) instrument. Finally, several of the other instruments on board can infer the presence of nearly-invisible clumps and clusters of centimeter-and-larger size ring particles by the depletions they create in Saturn's Van Allen belts of protons and electrons, and can even measure the faint breath of gaseous atmosphere enveloping the rings when the spacecraft approaches close enough (so far, only once in the mission, as described below and in Section 10.3).

The cameras give us the most detailed images of the rings in their full spatial glory; the spectrometers tell us the most about the ring composition, and the occultations give us the finest scale samples of ring structure (but only along linear cuts through the rings). It's left to the scientists to put all these bits and pieces together into a coherent story of the rings. Data have only recently started to flow to the science community, and this story is only beginning.

Launched in October 1997, Cassini spent its first two years in the inner solar system, during which time it encountered Venus twice and the Earth once to gather the extra orbital energy needed to make it all the way out to Saturn. Cassini passed Jupiter [12] in December 2000, getting a final gravity "slingshot" from the giant planet to send it outwards to Saturn. On July 1, 2004, after a nearly 7-year cruise, Cassini arrived at Saturn [13]. Plans went into action which had taken years to develop, in which commands occurring nearly every second of a complex, several hour long, series of events had to unfold in perfect sequence. The spacecraft crossed through the ring plane at a location deemed to be as safe as possible, and reoriented itself into the burn attitude. The burn commenced on time and the engines steered themselves along programmed paths to remain optimally oriented, as Cassini's trajectory was slowly deflected by Saturn's gravity (Figure 10.2). A "smart" algorithm monitored the 90-min burn on board, shutting off the engines when it determined that enough energy had been delivered to slow the spacecraft and allow Saturn to capture it into orbit. All this was done while the spacecraft was out of contact with Earth, with no possibility of ground-based intervention if anything went wrong, and while over a thousand scientists and engineers waited anxiously.

This period was the Saturn Orbit Insertion (SOI) period. The burn was planned to end just after Cassini's closest approach to Saturn. If it was successful, the spacecraft would know it (ground-based scientists still would not) and a carefully pre-planned

**Figure 10.2.** Cassini Saturn Orbit Insertion (SOI). Firing Cassini's main engine "backwards" slows the spacecraft down, letting Saturn capture it into orbit to begin its 4-year tour of the system (see Figure 10.3, color section). The spacecraft at this time is passing immediately above the huge ring system. After the 90-min burn, the spacecraft turned to make unique closeup observations, looking downward at the rings and sniffing their atmosphere of faint vapor.

series of unique science observations would be conducted. During this time the spacecraft would be seven times closer to the rings than at any other time in the mission, so it was an opportunity to conduct unique studies of the rings and the local environment so close to the planet. If something went wrong, the carefully planned science would be scrubbed and recovery activities would begin, to prevent Cassini from sailing right past Saturn and ending the mission. Everything went like clockwork, a credit to years of careful planning by JPL engineers.

Cassini entered orbit exactly as planned, beginning its 4-year tour of the Saturn system [14]. The tour is a series of elliptical orbits, all planned years in advance, which carries the spacecraft around to all corners of the Saturn system, allowing the planet, the rings, moons, and magnetosphere to be observed from a wide variety of angles and distances. Viewed from Saturn's north pole, the series of orbits looks like a flower (Figure 10.3a, color section), and indeed each revolution (or rev) is sometimes referred to as a petal. In Figure 10.3b (color section), we see how the tour petals look when viewed from Saturn's equatorial plane—many of them are inclined to the equator and extend both to the north and to the south. This is necessary to observe the rings, which lie in the equatorial plane, and to study both the north and south poles of the planet. This complex three-dimensional trajectory is made possible by frequent encounters with Saturn's large moon Titan—larger than several planets and itself a prime goal of study by Cassini (and of course Huygens). Passing close to Titan provides gravitational slingshot impulses which change the orbit trajectory significantly. Small corrections are then made using onboard fuel to be sure the next rendezvous with Titan occurs as planned, allowing the next orbit adjustment. All this takes very careful planning and celestial navigation, an astonishing feat by JPL's navigators and mission planners.

It took the assembled scientists of the project over two years of discussions and trade studies to select the specific configuration of petals that make up the tour, and another four years of weekly teleconferences by five groups working in parallel, with constant off-line preparation and homework, to debate and decide just what science observations would be the most critical to be done during every hour of every petal, and how much data would be allocated for storage of each onboard until the spacecraft could turn to Earth, about once a day, and send it down. This task was complicated because different instruments have different goals—often to look in different directions at the same time, and the total amount of onboard data storage is limited. Several times since the tour was decided on, adjustments have been required for various reasons, requiring scientists to scramble and rearrange the science accordingly. As of the publication of this book, the tour was about half complete in duration. One series of inclined orbits occurred in the spring and late summer of 2005, providing most of the ring observations described here; then the spacecraft went into about a year of elongated, equatorial orbits. The first of two final bursts of inclined orbits started in summer 2006. Ring observations, of course, account for only a fraction of Cassini's science return (see *http://saturn.jpl.nasa.gov* for some of the other spectacular results on the icy moons, the planet, and Titan).

## 10.3 MAIN RING STRUCTURE

**Radial structure** As mentioned above, one of the major uncertainties about Saturn's main rings has been: Just how much material do the thickest parts contain? For example, during Voyager, the tilt angle of the rings was only a few degrees as seen from Earth and the Sun, so neither sunlight nor Voyager's radio transmission could penetrate the densest regions, and the single observation of a star shining through the rings was limited by noise [15], so we have remained largely ignorant of its properties in many locations. Cassini was planned to arrive when the rings were tilted nearly wide open to the Sun and Earth for two reasons. First, sunlight is more easily able to penetrate the dense B ring, so images and spectra can be obtained from its unlit side. Second, the large opening angle relative to the Earth means that Cassini's radio occultations—passing its radio transmissions at three wavelengths through the rings—also can penetrate the thickest portions of the B ring. Figure 10.4 gives a sense of the very dense central core of the B ring. This same structure is revealed in

**Figure 10.4(a).** Image of Saturn's northern hemisphere, covered by the shadows of the main rings. (See next page for a different enhancement of this image.)

much greater detail by still-unpublished radio occultation observations. That is, the central region of the B ring is just as opaque to the probing radio waves, and the same slightly more transparent channels are seen within and around it. The meaning of this structure will need to await further analysis.

We can also look directly at the unlit face of the rings as revealed in Cassini images, such as shown in Figure 10.5. This image was taken of the north (unlit) face of the rings, and shows how the dense B ring blocks most of the sunlight, which at this time was falling on its southern face, from getting through. The moderate optical depth A ring is of intermediate brightness, and the least optically thick rings—the C ring and Cassini Division—appear the brightest because they are optically thin enough for us to see through them to the lit particles on the southern face.

**Spiral density and bending waves** One of Voyager's most important discoveries about the rings (see Chapter 9) was how moons create spiral waves which propagate slowly

**Figure 10.4(b).** A brightness-enhanced and stretched view of the same image, showing a very dense central core of material in the B ring, which seems to be cut by a few lower density channels.

through the rings. Each moon has a well-defined orbit period—the time it takes to orbit Saturn. In the rings, the orbit period changes continuously with distance from the planet; material closer to the planet orbits in a shorter time. Thus, there is a large, but countable, number of places where the ring-particle orbit periods are a simple integer fraction of some moon's period; these locations are called orbit resonances [16]. At resonances, the small gravitational forces from a moon act repeatedly on the same ring material, and after some number of orbits cause noticeable disturbances in its behavior. These disturbances create condensations of ring material, and the tiny gravitational forces from these condensations generate further disturbances in particle motions which propagate away from the resonance in a tightly coiled spiral that looks like a watchspring (Figure 10.6). Spiral density waves, where the wavelength decreases outwards, are common; moreover, for moons which are also *inclined* relative to the rings, the resulting up-and-down forcing creates spiral waves of vertical motion—corrugated or flapping structures called spiral bending waves, where the wavelength decreases inwards (Figure 10.6 shows examples of both). The wave patterns are fixed

**Figure 10.5.** Saturn's main rings from the northern (unlit) face. The dense B ring appears the darkest. The A ring, at left, contains two empty gaps; the most noticeable is the Encke gap. The Cassini Division lies between the A ring and the very dark, and dense, B ring. Its outer portion is brighter in this view, partly because the particles there are more reflective than those in the inner Cassini Division. The outermost part of the C ring is visible to the right, echoing the appearance of the Cassini division.

in the frame of reference rotating with the moon which causes them, so ring material, which moves more quickly than the moons, actually moves through these wave crests and troughs. The spirals can have multiple arms, depending on the order of the resonance (the number of orbits of ring and moon needed for them to become aligned again).

Much of the theory of spiral waves is in fairly good shape (Section 9.5), so their observed properties can be used as powerful tools to measure the properties of both the moon that causes them and the local ring material through which they move. For instance, the local wavelength, or distance between crests, is a measure of the local surface mass density of the rings (grams per square centimeter), which tells us the total mass of the rings. Knowing both this property, and the optical depth of the rings, tells us the typical particle size (because larger particles have more mass per unit area). Or, if we have independent information on the particle size—for instance, from a radio occultation—we can infer the density of the particles and thus something about their composition. Moreover, the amplitude of the waves (their contrast or strength, from

**Figure 10.6.** Spiral density and bending waves in Saturn's rings. Each wave is caused by a single orbital resonance with a different moon. Here, the bright, broad wavetrains are caused by Mimas—a density wave at right center and a bending wave at left. The next smaller wavetrains are caused by Pandora and Prometheus. The finely spaced lines throughout are each individual wavetrains—here not resolved—caused by Pan in the nearby Encke gap, which is much closer and much smaller. In addition to the images, dozens of stellar and radio occultations profile the structure of these waves at different longitudes.

crest to trough) tells us the mass of the moon that causes them. From this and images of the moon which give us its volume, we can determine the density of the moon (Section 10.5).

Density waves damp out as they propagate, because their energy of motion is dissipated in collisions between particles at the small random velocities discussed later in this section. The damping length of a wavetrain can thus tell us something about the

**Figure 10.7(a).** Closeup images of spiral density waves (above and left part of 10.7(b)) and spiral bending waves (right half of 10.7(b)) in Saturn's A ring, taken by the Cassini ISS instrument during the short SOI time period when the spacecraft was flying immediately above the rings. The images are small sections of spirals that are as tightly wrapped as watchsprings. Because these are taken from the unlit face of the rings, opaque regions appear darker. The image on the opposite page—Figure 10.7(b)—is a higher resolution piece of the terrain covered by Figure 10.6; the image above shows a grainy structure dubbed "straw".

local random velocities, or collisional velocities, between ring particles. Both stellar occultations by VIMS and UVIS, and radio occultations by RSS—all still under analysis—have revealed perhaps hundreds of new spiral waves and have traced the damping of many of them much more carefully than possible with Voyager observa-

**Figure 10.7(b)**

tions [17]. Some waves are now seen to propagate more than 600 km—much farther than expected and consistent with predictions of theories that many scientists had previously treated with skepticism. Two extreme closeups of spiral waves are shown in Figure 10.7(b), taken by the Cassini ISS team just after the SOI burn, having a spatial resolution of only 200 m/pixel!

One interesting new observation of interest revealed in Figure 10.7(a) is the speckled dark blotches seen at the lower left, in the troughs (bright, low-density valleys) between the opaque, dense crests of density maxima. These appear to be densely packed clumps of particles, jammed and stuck together as particle trajectories converge in the crests of density waves. Emerging from the dense crests, these plugs of ring material—as long as a freight train and thicker in diameter—eventually scatter apart as they collide with other particles. Understanding this effect will help us understand the physics that transpires in the dense crests of density waves.

**Main ring irregular structure** Given the complexity of the irregular structure [18] and the variety of data required to unravel it, much of which is still under analysis, it's a little early for specific advances to be evident. Theoretical advances in this general area have been substantial since Voyager, however. A number of studies have converged on a process called the "pulsational instability" or "overstability" as an explanation for irregularly spaced, fine-scale structure in the rings [19]. This process relies on a combination of viscosity and self-gravity of the ring material, acting in opposition. Self-gravity tends to make dense structures become denser, depleting surrounding regions, but if viscosity increases with increasing density faster than gravity does, viscous diffusion spreads the dense material out again, causing the *adjacent* regions to become denser in their turn. A number of theoretical models show that very fine scale structure can indeed develop if the optical depth is sufficiently high. The length-scale of the structure is very fine—some tens of ring thicknesses in size, or perhaps a few hundred meters. Also, the models show a transient structure, sloshing back and forth without ever forming a permanent, fixed pattern—but always present. In fact, Voyager observations (Chapter 9) had already shown that the finest scale structure in the B ring—that seen at its outer edge—was not azimuthally symmetric but varied with longitude. Whether this longitudinal variation also implies temporal variation is still not known, and more observations and analysis are needed. It has been suggested that, over time, larger-scale structures might grow or "anneal" from the fine-scale structure, but models show that the fine-scale structure always remains underneath.

Cassini observations don't confirm all these predictions, but there are some interesting correspondences. Figure 10.8 compares two moderately high-resolution images of the lit face of the B ring, showing the irregular structure in much better detail and higher sensitivity than seen by Voyager. The images have the same resolution, and it is readily apparent that there is abundantly more very fine scale structure near the outer edge of the B ring than in its center.

Moreover, Cassini ISS images taken at SOI, having a resolution of 200 m/pixel, do show fine-scale structure in abundance, in several places—and preferentially where the ring optical depth is the largest (Figure 10.9, two right panels). At locations only a few hundred km away (Figure 10.9, two left panels), the familiar irregular structure is seen with a scale of 100 km or more, which entirely lacks this ultrafine structure. So, in some ways the data are in accord with theoretical predictions that viscous overstability (pulsational instability) will occur in only the most optically thick regions; however, we still seem to need a different mechanism to generate the longer (few hundred km) scale irregular structure that permeates the B and inner A rings where the optical depth is only a little bit lower. Furthermore, we need to understand why the optically thickest regions of all—in the middle of the B ring—don't seem to show as much of this fine-scale structure as regions in the outer B ring and inner A ring, where the optical depth is actually smaller. Cassini radio occultation data and stellar occultation data will ultimately contribute heavily to unraveling this story, providing higher spatial resolution and multiple cuts across the rings at different longitudes and times. However, these data remain in the early stage of analysis.

**Figure 10.8.** Two images (this and next page) of the lit face of the B ring, taken at nearly the same time and with the same spatial resolution, showing its irregular structure. In the middle B ring (this page) we see abundant structure with scale of a few hundred km, but little if any very-fine-scale structure. In the outermost B ring (next page), terminating with the empty Huygens gap, its ringlet, and the bands of the Cassini Division, abundant ultrafine scale structure is seen. The lower images on both pages are enlargements of about a quarter-scale central part of the upper ones, again both at the same scale.

**Figure 10.8** (*cont.*). Outer B ring. Note the abundant fine-scale structure compared with the middle B ring shown on the previous page.

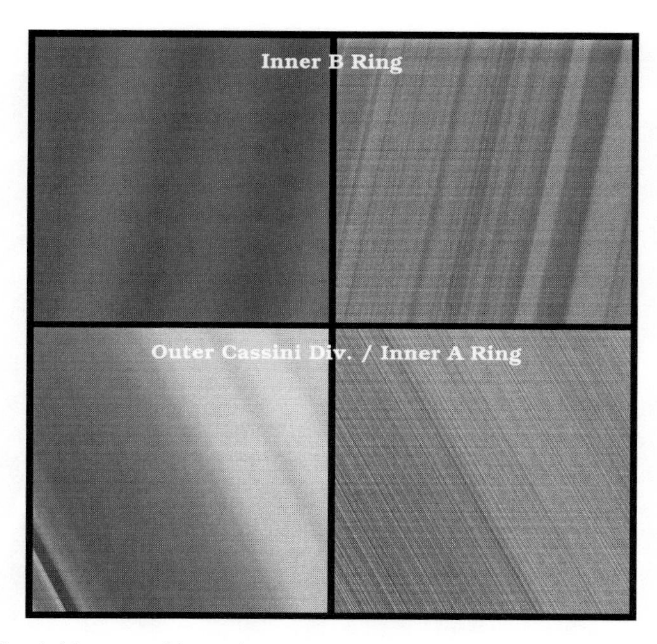


**Figure 10.9.** Cassini images of irregular structure in Saturn's B and A rings, taken during the SOI period. These images have a resolution of 200 m/pixel and show the rings from their unlit face, so darker regions tend to be more opaque. In the upper two panels, we see two regions in the inner B ring—one where the structure has a smooth, 80-km scale, with no trace of fine-scale structure, and the other where km-scale structure is everywhere. Similar fine structure with km scale is seen in the inner A ring, which is also optically thick, but not in nearby regions.

**C ring and Cassini Division** The C ring, Saturn's innermost ring, is an enigma of another sort. In many—if not most—ways, the C ring most closely resembles the Cassini Division—which is not at all empty but a ring, of sorts, in its own right. Their similarities are evident in plots of their optical depth as a function of radius (Chapter 9). Both regions contain several empty gaps, and otherwise empty gaps containing eccentric, sharp-edged ringlets. A few of the empty gaps in the C ring are identified with known satellite resonances, but just as many are not. The C ring also contains several fine examples of spiral density waves which have no assigned cause [20]. Moreover, the C ring presents us with a quite beautiful and simple, but completely unexplained, series of moderate optical depth "plateau" features that are nearly regularly spaced and nearly symmetrically placed about its outer (unexplained) Maxwell gap (Figure 10.10). This structure is also evident in Figure 10.19 [21]. Cassini has obtained a number of images of these regions, has taken several surveys for small embedded moonlets (none have as yet been found), and is starting to do full azimuthal studies to search for telltale structure that might provide evidence for moonlets too small to see directly. Radio and stellar occultations have also begun to sample the region, again with no specific conclusions as yet released regarding the cause for this intriguing structure.

**Figure 10.10.** The outer C ring, surrounding the Maxwell gap with its embedded, eccentric ringlet (right center). The bright plateau features are not surrounded by gaps or set off in any way, yet maintain their moderately sharp edges in spite of the expected smoothing by diffusion and spreading. They are about 300 km wide. There is no known resonant structure to explain their observed symmetry about the Maxwell gap.

**Main ring azimuthal structure** As described in Chapter 9, the A ring has long been known to display an azimuthal asymmetry which is thought to be caused by the varying cross-sections of elongated structures, caused by ongoing gravitational instabilities, which are angled to the line of sight. Cassini's many stellar and radio occultations, and even the structure in the ring temperature variations, are showing these differences in wonderful new detail, although no results have yet been published. Figure 10.11 illustrates how this works. Occultation traces passing through the rings at different longitudes result in different apparent optical depths, because the ring structure is not uniform. As more occultation traces are made at different longitudes and elevation angles, it is becoming possible to model the horizontal and even the vertical extent of these structures, and ultimately more will be learned about the particle size distributions which make them up as well [22].

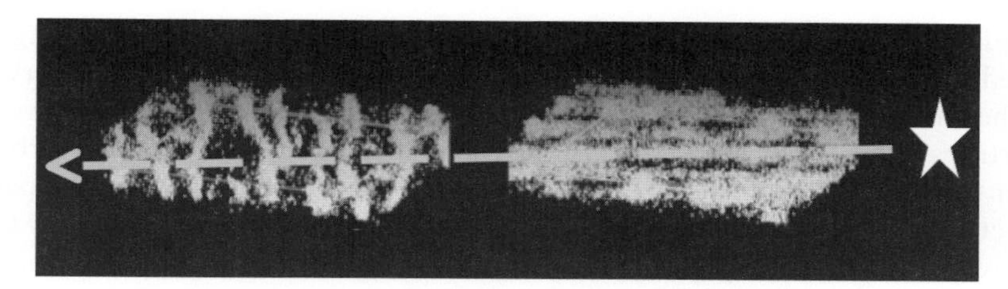


**Figure 10.11.** Schematic of a stellar (or radio) occultation trace through a patch of rings in which stranded gravitational instabilities, or gravity wakes, have formed. The wakes have a constant typical pitch angle relative to the direction to Saturn, and thus their angle relative to the observer depends on their longitude. Occultation traces passing through longitudes as seen at left let more light pass through, and appear to have lower optical depth, than occultation traces through regions appearing as at right, where more of the starlight is blocked.

**Main ring vertical structure; ring volume density** One of the remarkable things about Saturn's rings is how thin they are vertically for their enormous radial extent (Chapter 9, Figure 10.12). Ever since their discovery, the fact that they essentially vanish when seen edge-on from Earth has fascinated observers. For the last few decades it has been

**Figure 10.12.** Where'd they go? Although nearly as far from one side to the other as from the Earth to the Moon, Saturn's rings are locally only a few tens of meters thick. Even what we see here, imaged from Cassini in an equatorial orbit, is light scattered from rippling vertical structure in the rings—spiral bending waves a few km thick—and vertically inclined rings such as the F ring. At the top are the shadows of the main rings on Saturn's northern (winter) hemisphere. The blackest shadow is from the densest (B) ring. The upper shadow is from the A ring, showing light coming through the Encke gap.

realized that even other forms of vertical structure in the rings—spiral bending waves, inclined ringlets, and so on—mask the local ring thickness in edge-on observations.

The true local thickness of the rings is determined by the small vertical velocities of particles—the larger these are, the higher particles wander in their orbits, the thicker the rings are, and the lower is the volume or packing density of the particles. Theoretical models which include inelastic collisions between particles imply that the rings—especially the opaque rings where collisions are common—can't be more than a few times the size of the largest particles in vertical thickness, and thus that the packing density of particles can't be extremely low.

For decades, these dynamical expectations have been hard to reconcile with the opposition effect of the rings (Chapter 9) in which the ring brightness increases sharply when the Sun and Earth line up and sunlight is scattered directly backwards to the Earth (also known as the zero-phase effect). This was originally attributed to a shadow-hiding process, and implied that the particles were widely separated in a low-volume density ring. Another theory emerged, based on laboratory observations, in which the opposition effect is caused by interference of photons which travel along opposite but identical trajectories, which implies that shadow hiding plays no role and thus that the ring volume density does not need to be extremely low [23]. The predictions of these two theories were quite different regarding how the effect should vary with the reflectivity of the ring particles. Cassini observations of this effect were planned in order to resolve this difficulty. Figure 10.13 illustrates these observations. In Figure 10.13, we see a Cassini ISS image of the rings taken by the spacecraft with the Sun immediately behind it, so the light is directly scattered backwards. This geometry allows photons from the Sun to traverse slightly different paths having the same total distance between the rings and the camera, allowing them to add coherently together when the phase angle (the angle between the Sun and the camera as seen from the rings) becomes extremely small (i.e., at the bright spot). The test of this effect was whether it would behave the same for particles of high and low albedo, and for rings of high and low optical depth.

The VIMS team observed the same geometry simultaneously, and while their spatial resolution is not as good as that of Cassini's cameras, they have a broader coverage in wavelength and particle reflectivity which has allowed them to conclude that hiding of shadows is not, after all, responsible for this effect and, instead, coherent backscattering apparently is ([24]). This reconciles the observations with dynamical expectations, but leaves us with one less way to measure the actual volume density of the rings.

**Diffuse rings** Saturn's diffuse, outlying rings have never been in the spotlight, lacking the exotic structure of the main rings. Yet, they are of profound importance because Cassini spends a considerable time simply crossing through them, and we need to understand the nature and abundance of their particles. Furthermore, these modest structures have become ever more interesting because of their time variations.

*The E ring*: This, for example, is dominated by tiny and short-lived grains; this aspect, plus its association with the unusual moon Enceladus, which Voyager showed to have flow-like structures on its surface, were clues that something perhaps unusual

**Figure 10.13.** The "opposition effect" is apparent in this Cassini ISS image of the main rings, with the C ring at lower left and the Cassini Division and A ring at the top right. The Sun is directly behind the spacecraft, and the bright spot in the B ring is the directly backscattered light from the Sun. Its small angular size (less than a degree) is an indication of "coherent backscattering" in the grains of the ring particles rather than a measure of their spacing. Images like this were taken to track this bright spot as it moves across the face of the entire ring system. When their analysis is complete we will have a better understanding of exactly how the opposition effect varies from one location to another.

was occurring in and around this ring (Chapter 9). Cassini found, months before even getting to Saturn, that the E ring had new surprises in store. The UVIS instrument routinely maps the entire Saturn system for fluorescence from atoms which might be degassed from various rings or moons. Between the end of 2003 and March 2004, the UVIS instrument saw a huge increase in the abundance of atomic oxygen in and around the E ring ([25], Figure 10.14). Now that Cassini has discovered active jets and plumes of dust and ice being degassed from the near-surface of Enceladus [26], it becomes easier to understand such time variation as merely a consequence of sporadic eruptions from Enceladus. This is an idea that actually goes back several decades, but—in the absence of any hard observations of Enceladus—was regarded as somewhat speculative at the time [27]. Other odd and unconfirmed features have been seen

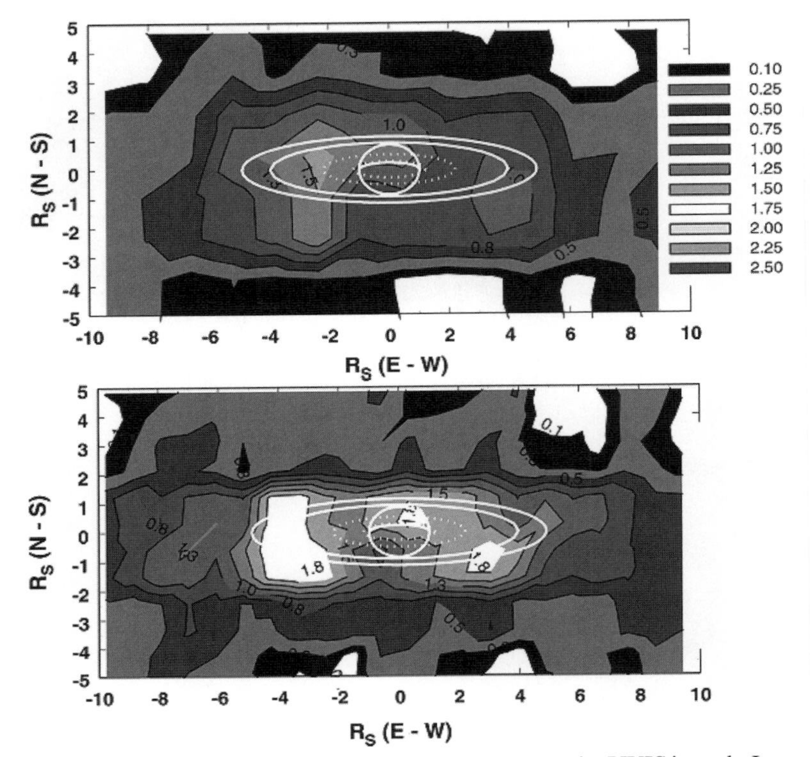


**Figure 10.14.** Abundance of atomic oxygen in the Saturn system by UVIS in early January 2004 (top) and in February 2004 (bottom). The band outlined in white is the E-ring core. The amount of oxygen that appeared in this one month (and then disappeared again in the next month) is comparable with the entire mass of all the known E-ring micron-sized grains. This puzzle has apparently been resolved by Cassini's discovery of active degassing of water from the moon Enceladus, which orbits in the core of the E ring and presumably supplies its micron-sized ice grains.

in or near the orbit of the E ring [28]; perhaps Cassini will be able to confirm such features during its 4-year tour.

*G ring arc*: Orbiting between the huge, broad E ring and the main rings is the G ring—thought to be rubble left over from disruption of a former moonlet [29]. Actually discovered by Pioneer 11 from its influence on Saturn's Van Allen belts, it was first imaged by Voyager (Chapters 2 and 9). Cassini has been watching the G ring carefully prior to encounter, as scientists and mission planners carefully chose and monitored a trajectory for Cassini to avoid being sandblasted by ring particles on our first approach to Saturn. Since that time, the G ring has been observed in a number of geometries. It remains a diffuse band of material, in about the same place as seen by Voyager, but seems to have given birth to a new feature—an azimuthally incomplete arc of debris, concentrated at certain longitudes (Figure 10.15). Probably, whatever still-unseen moonlets remaining within the G ring's diffuse core—providing

**Figure 10.15.** Saturn's G ring is several thousand km wide, but its core is primarily visible here. Note in this time sequence an extended clump or arc, starting at bottom center in the left image, moving to the ring ansa or tip in the middle image, and continuing to the far arm in the right image. This feature has been tracked in a number of other images.

material to replenish the ring by constant micrometeoroid erosion—can also collide or be hit with a big enough meteoroid to throw off large amounts of debris. The orbital rate of the arc is not yet well enough known to say much about its dynamics. It will be of interest to see how long it lives and how it spreads.

*The D ring*: Lying even closer to Saturn than the C ring, the D ring was glimpsed only once during each Voyager flyby. It is composed of a number of faint ringlets, each somewhat reminiscent of the G ring (Figure 10.16). However, Cassini observations

**Figure 10.16.** Comparison of structure in the D ring as seen by Voyager and Cassini. The bright material in the lower image is the inner edge of the C ring. Two of the D ringlets align, but its formerly brightest one has either disappeared or moved and been replaced by a relatively low brightness feature. The insert shows a new kind of very regular pattern seen at the very outermost edge of the D ring.

showed that one of these ringlets has either changed location or vanished entirely to be replaced by another nearby fainter ringlet, or perhaps by a gap, even while others seem to be found in the same place. Moreover, a new kind of regular pattern, which has no explanation, has been seen at the very outside edge of the D ring, just inside the C ring.

The particles making up the D ringlets don't contain much mass, so it's not implausible for some impact to have simply blown apart a big boulder into rubble that has spread and become eroded to produce dust; however, understanding how the previously dominant ringlet could just vanish so quickly is more difficult. Modeling of absorption by local boulders is needed to see if it can remove dust in the two decades since Voyager, while remaining invisible today.

## 10.4 RING COMPOSITION AND PARTICLE PROPERTIES

**Ring composition** The composition of the rings is one of their oldest puzzles, and perhaps the most important for establishing their origin (Section 10.7). It has long been known that water ice is abundant in the rings, from near-infrared spectra and microwave observations (Chapter 9). Cassini's VIMS instrument measures spectra at wavelengths from visible through 5 micrometers in the near-infrared; it is at these near-infrared wavelengths where the strongest absorptions due to water ice are found. These observations, most of them still under analysis, clearly show the diagnostic water-ice bands, now in much greater detail than before and with spatial resolution that can show how the ring composition varies with location. In Figure 10.17 [30], we clearly see the diagnostic water-ice absorption bands at 1.5, 2.0, and 3.0 micrometers. Laboratory analysis of water ice at low temperatures indicates that pure water ice has a spectrum that looks very similar to the A ring (Figure 10.17a). That is, another weak water absorption feature appears at 1.06 micrometers, and the brightness continues to rise towards shorter wavelengths. However, notice that the other three spectra in Figure 10.17, while also dominated by strong water absorptions, differ at the short wavelengths, with the 1.06-micrometer band disappearing and the spectrum sloping downwards from 1.4 micrometers towards shorter wavelengths. The effect is especially noticeable in the Cassini Division and C ring, regions which had been known to have darker and less-red colors at visible wavelengths (see below). While much more analysis needs to be done, initial indications are that this distinctive spectral shape is evidence for iron, probably in the form of silicates, being more abundant in the Cassini Division and C ring than in the A or B rings. The F ring shows evidence for a carbon–nitrogen-based organic material similar to that seen by Cassini's VIMS instrument on Saturn's irregular moon Phoebe, which is a very dark, comet-like object—apparently a refugee from the Kuiper Belt, an ensemble of primitive comet-like objects orbiting outside Neptune [31]. Care must be taken with this sort of analysis, because scattered light from Saturn, having its own spectral properties, can confuse the interpretation. For example, Figure 10.18 (color section) shows a VIMS map of Saturn where the planet's telltale methane absorption features have been colored a reddish pink. The intrinsic ring spectrum at longitudes in the rings which are illuminated by the hulking bright planet (left side) or where sunlight traversing the upper atmosphere of the planet along the edge of the shadow is corrupted by atmospheric absorption (right side, along edges of shadow).

Figure 10.17(c) shows how the strengths of these water-ice absorption features vary with radius near the inner edge of the B ring, which tells us how the ring composition and/or surface grain size varies with location. This variation can be compared with Figure 10.17(b), which shows the ring structure measured by VIMS as a function of radius. These observations (in the SOI time period), showed the unlit face of the rings, thus the inner B ring (outwards of 92,000 km) appears to be about the same brightness as the C ring. Notice that although the ring brightness varies dramatically from place to place as the amount of material varies (middle panel), the strength of the water-ice features varies only smoothly (lower panel) and, in particular, there is no dramatic change at the boundary between the B ring and C ring. This indicates the composition varies smoothly across this very dramatic boundary in optical depth. Small offsets appear to be associated with the C ring plateaus.

The new color images obtained by Cassini also contribute to this story. It has been known for about a decade, from Voyager and Hubble Space Telescope observations, that the spectrum of the rings at visible wavelengths rises steeply from the blue to green range, and remains flat out through the red, giving them a tan, taupe, color [32]. That is, they are not white, as pure water ice would be. Notice, for example, Figure 10.19 (color section), where Cassini captured both the rings and the nearly pure water-ice surface of Enceladus in the same image. Even to the naked eye, the difference is noticeable. Whatever it is that gives the rings this tan color can't be a major constituent, but it doesn't take much colored material mixed into pure ice to give it a detectable color. Some work has suggested that reddish organic molecules, common in the outer solar system, give the rings their reddish hue. Neptune's large, icy, probably captured moon Triton, for example, appears quite pink [33]. Careful inspection of the full color mosaic in Figure 10.20 (color section) reveals some areas in the rings which look less red than others. Analysis of Voyager and HST observations has previously shown that the ring color varies smoothly across ring boundaries [10], much like VIMS has now shown the strength of water-ice absorption to do.

The Cassini UVIS instrument has also mapped the ring spectrum in the ultraviolet, where water has a strong UV absorption edge. Their measurements of the reflectivity of the rings tend to show the ring brightness increasing outwards all through the A ring, which might imply brighter ring particles [25], but must be considered together with inferences from VIMS data where the outermost piece of the A ring appears to be richer in the iron-based silicate "mystery material". This is thus an interesting situation, which will require more sophisticated modeling of both observation sets, including the effects of ring-particle packing density and surface grain size on reflectivity in the two different wavelength regimes.

**Ring particle sizes** A perennial question, as discussed in Chapter 9, has been "what are the sizes of the particles in the rings?" The inability to detect the rings in microwave emission during the 1960s and early 1970s led to a belief that the particles were microscopic—too small to emit radio waves. However, their strong radar return ruled that possibility out, and the combination of these two observations implied that they

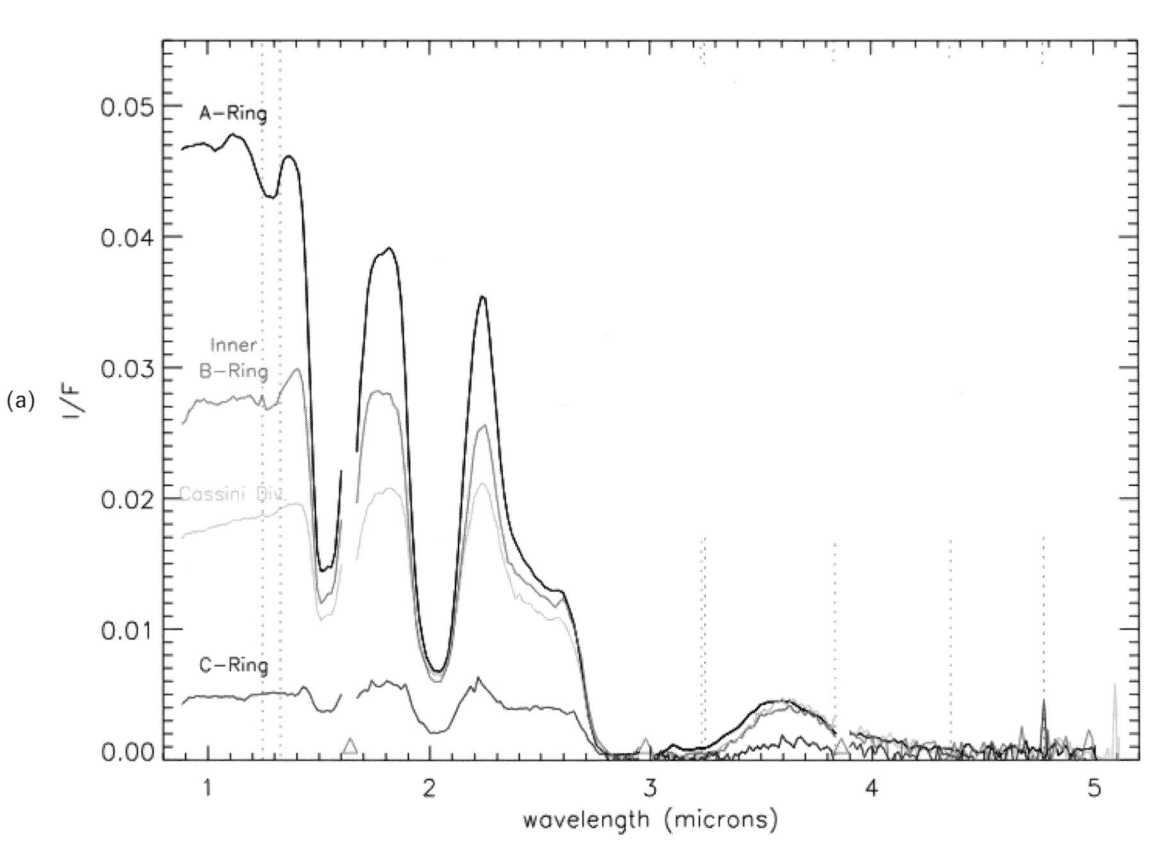


**Figure 10.17.** Cassini VIMS observations of Saturn's rings. (a) Near-infrared spectra of different regions, showing the prominent water-ice absorption bands at 1.5, 2.0, and 3.0 micrometer wavelength. Vertical dashed lines indicate "hot pixels" which can provide spurious results. The presence of some non-icy, probably iron–silicate absorber, is indicated by brightness decreasing towards wavelengths shorter than 1.4 microns, and the absence of the weak 1.06-micrometer water ice band.

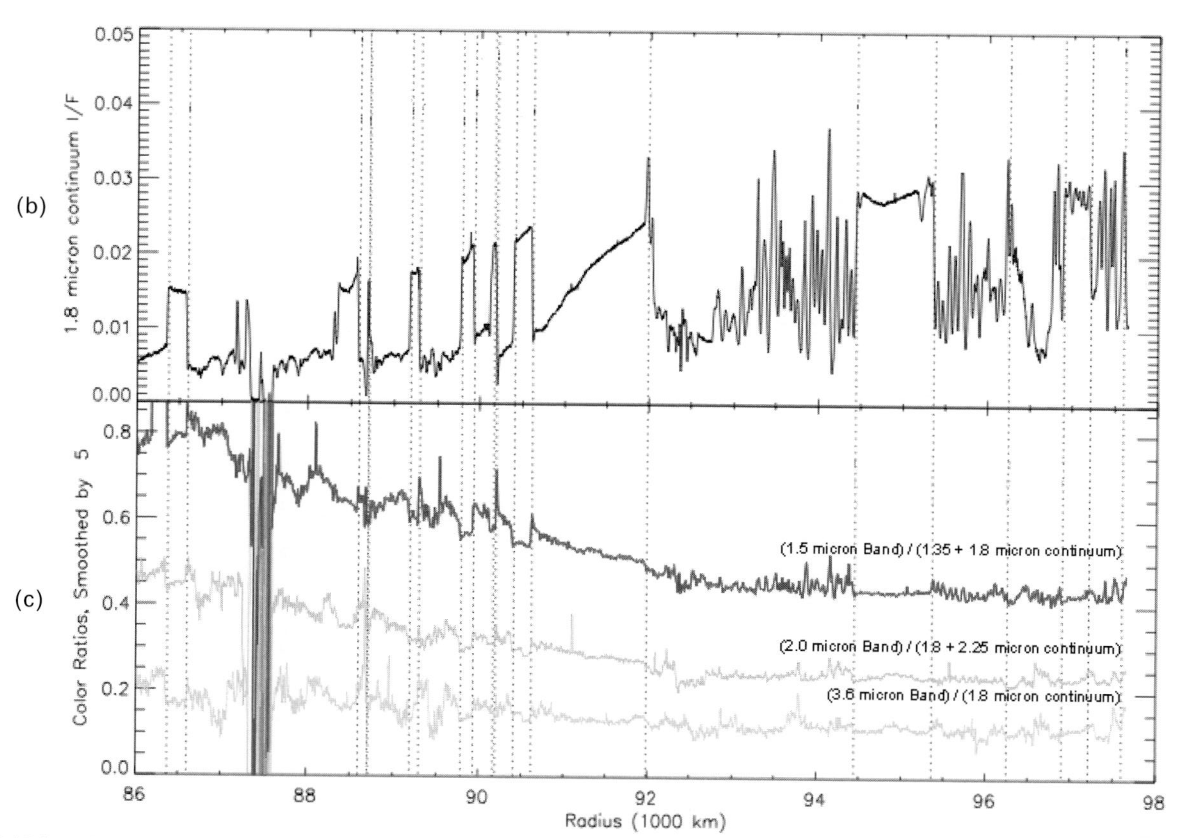


**Figure 10.17** (*cont.*). Cassini VIMS observations of Saturn's rings. (b) Radial profile of ring brightness from SOI observations of the unlit face of the rings, with the abrupt inner B-ring edge just outside 92,000 km. Note the six rectangular "plateaus" of moderate optical depth in the C ring, between 86,000 and 91,000 km, as seen in reflected light in Figure 10.10. (c) Radial trends of water-ice band strengths. Note the lack of a significant jump at the B–C ring boundary, and only small discontinuities across the plateaus in the outer C ring; instead, a generally smooth trend is seen over several thousand km.

had to be somewhat on the order of a radio wavelength in size—tens of centimeters to a meter. The great value of radio observations of the rings derives from this property—that the scattering behavior of particles at wavelengths close to their size is strongly variable and highly diagnostic. The relative abundance of particles between a centimeter and 10 meters in radius is a key aspect of local ring dynamics and the evolution of ring composition (Sections 10.3 and 10.6).

Cassini's Radio Science Subsystem (RSS) observations of the rings were designed to provide this information on a fine radial scale. The main antenna transmits pure tones at three different radio wavelengths (2-, 4-, and 13-cm wavelength) to Earth when the spacecraft goes behind the rings, or is occulted by them. These radio occultations also study the planet's ionosphere and atmosphere. Small particles block or scatter the shortest radio waves better than the longest waves, so *relative* differences in radio opacity between wavelengths can be interpreted as particle size differences. To date, 12 separate cuts through the rings have been successfully observed, each at three radio wavelengths. Sample results from just one of these are seen in Figures 10.21 and 10.22 (color section), where a single occultation trace through the rings at one longitude has been stretched around to resemble a ring; the radial variations and structure are real but only at one longitude. In these figures, the opacity of the rings at the longest wavelength is colored red and that at the shortest wavelength is colored blue. Thus, regions appearing red have the most abundant large particles, and regions appearing blue have more small particles. Figure 10.21 shows a global view of the entire main ring system, the white bands represent the dense core of the B ring, which was opaque to radio waves at two of the three wavelengths, so no size information is yet obtainable; further analysis is underway. Notice the similarity of this white (opaque) region to the global view of the densest ring regions seen in their shadow (Figure 10.4). Also notice how the C ring, Cassini Division, and outer parts of the A ring (green- and blue-colored) have the smaller particles. Figure 10.22 focuses on the Cassini Division and A ring; notice how the particle size distribution changes from place to place, with small particles becoming especially dominant towards the outer edge of the rings (increasingly blue color in this figure).

**Rings in the thermal infrared** Cassini's CIRS instrument measures thermal infrared emission from the rings across a wide range of wavelengths. At the shorter wavelengths of tens of microns, the grainy particle surface is a good emitter and the actual temperature of the particle can be directly measured. Figure 10.23 shows how the ring-particle temperature varies with location in the rings in several different geometries—lit and unlit faces, at low and high phase angles [34]. At a given phase angle, the unlit face is, naturally, cooler than the lit face—especially for the dense B ring.

Note also how the C ring and Cassini Division are noticeably warmer than the A and B rings (at low phase angle). This testifies to the fact that the C ring and Cassini Division particles are darker. The smooth bumps and dips in the CIRS data, which illustrate its fairly low spatial resolution, are connected to large-scale variations in the ring optical depth in ways that remain to be modeled, but will help us understand the dynamics of how particles move from one face of the rings to the other. At high phase angles, there is little difference between the particle temperatures in the C ring and

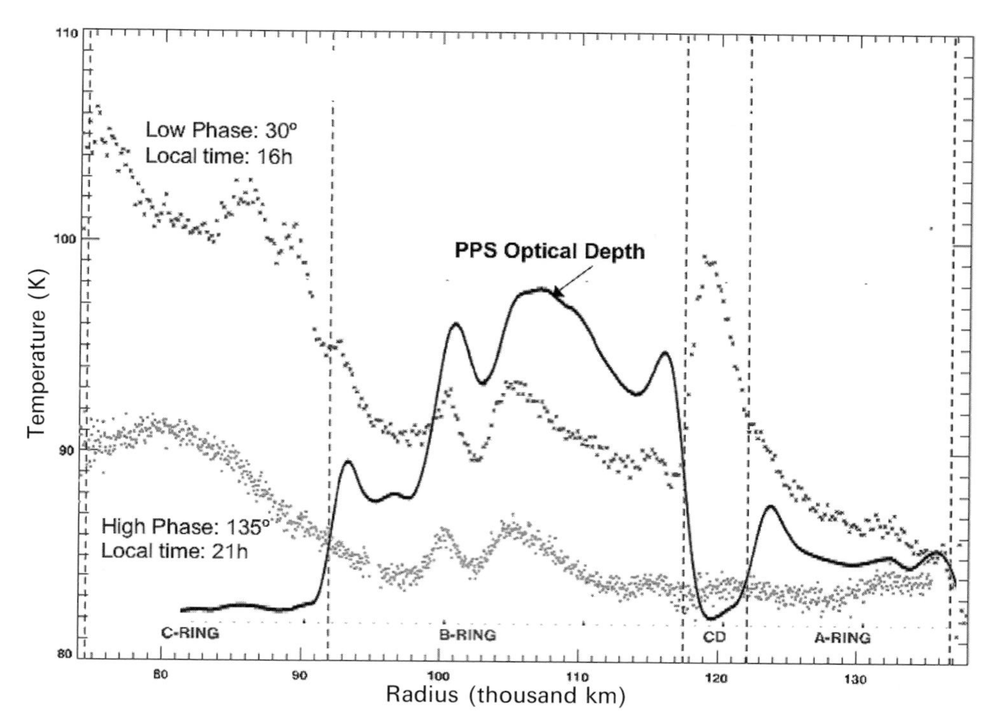


**Figure 10.23.** Radial profiles of ring particle temperature from CIRS, showing temperatures on the lit face in two different observing geometries. The spatial resolution is low and blurs sharp ring boundaries (shown as vertical dashed lines). The C ring and Cassini Division appear warmer than the B ring and A ring in low phase angle observations (where the "lit" faces of their particles are being observed), but the Cassini Division shows no thermal contrast at high phase angles (where the "dark" faces of ring particles are being observed). Also shown (solid curve) is the ring optical depth from the Voyager stellar occultation experiment, smoothed to the CIRS resolution. The most optically thick rings—which are usually the brightest in reflected light observations—surprisingly have higher temperatures. The reverse is seen on the unlit face (not shown) where particles in the most optically thick regions have the lowest temperatures. This result is hinting at some dynamical control of particle temperature; for instance, particles might get "trapped" on one face of the rings—becoming unusually warm on the lit face and unusually cold on the unlit face.

Cassini Division, and those of the A and B rings; this seems to imply that the particles spin at rates not too different from their orbit periods. The fact that the A ring shows the least variation across geometries might suggest that the most rapidly rotating particles are found there. Observations of this nature will continue as the Sun sets on the rings, giving us more insights into their vertical structure and local dynamics.

Another observation that comes from the CIRS instrument is the emissivity of the particles. At CIRS's longest wavelengths of 0.5–1 mm, the grainy ring-particle surfaces are no longer guaranteed to be good emitters. We already know from the low microwave brightness temperatures of the rings at centimeter wavelengths that the

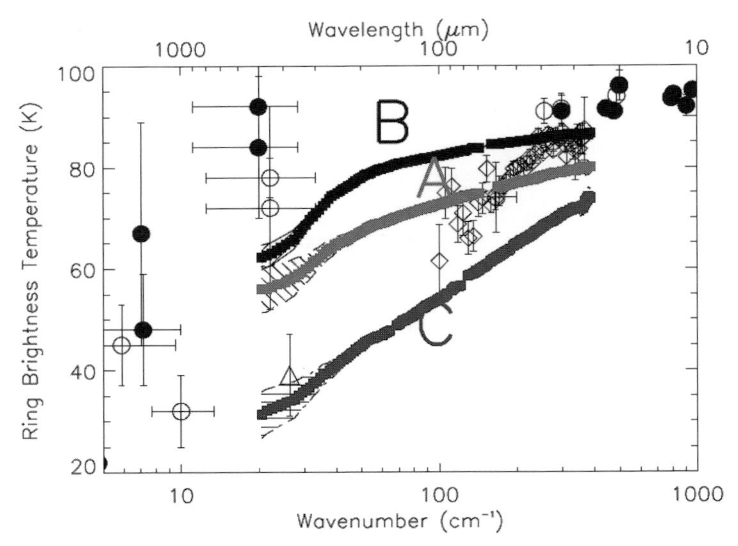


**Figure 10.24.** Prior (and inconsistent) ground-based observations of ring brightness temperature (points; see Chapter 9) with new Cassini observations (smooth profiles for A, B, and C rings, with uncertainty ranges at long wavelength end). Brightness temperature measures the emissivity of the rings, which decreases to longer wavelengths because the icy particles are becoming more transparent and better scatterers. The different shapes of the profiles for the three main rings contain information about the different surface properties of the particles; as the mission progresses, the wavelength range will extend towards 1 mm.

emissivity is very low there (Chapter 9); the transition regime is sensitive both to grain size on the surface of the ring particles, and to the abundance of the non-icy component of the ring material. Figure 10.24 is a very preliminary result from the first few observations of this type [35]. Understanding the meaning of the emissivity drop towards longer wavelengths (here plotted as ring brightness temperature) will require more sophisticated models of the ring-particle surfaces. Because CIRS observations of this type, and VIMS observations of the depths of the water-ice features, both depend on surface grain size in different ways, it will be possible to distinguish between grain size effects and compositional effects. Because CIRS operates at a longer wavelength than VIMS, it samples more deeply beneath the ring-particle surfaces to see how the abundance of the non-icy material varies with location.

**The ring atmosphere** Even though most of Cassini's observations of the rings must be made remotely, the period near SOI did provide a brief opportunity to sense the ring material *in situ*, even if indirectly. Cassini carries several instruments designed primarily to sample the magnetosphere of Saturn and measure the abundances of different elements in it. As Cassini skimmed across the unlit face of the rings, just after the main burn was complete, a number of these observations were made. It had been expected that the ring atmosphere would be primarily breakdown products of water—

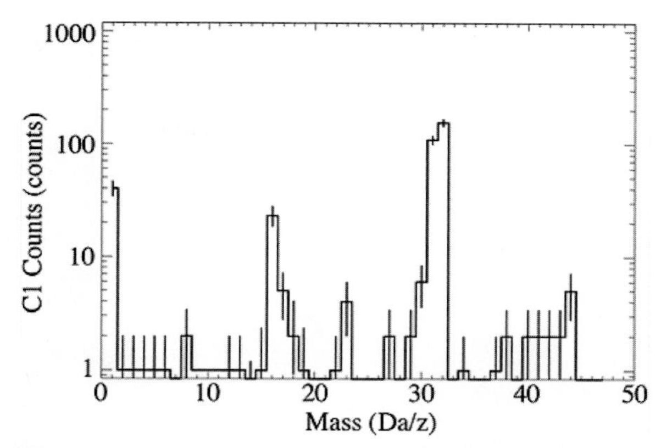


**Figure 10.25.** Results from the Cassini INMS instrument showing the composition of the ring atmosphere. Major peaks at 16 and 32 Da/z (atomic mass units or a.m.u. per charge) indicate O and $O_2$, respectively. Sodium appears at 23 a.m.u. and perhaps $CO_2$ at 44 a.m.u. Other products are H at 1 and OH at 17 a.m.u.

H and OH—but instead, molecular ($O_2$) and atomic (O) oxygen were by far the most abundant materials, and sodium was also present (Figure 10.25 [36]). Models of the environment have been developed which explain this interesting result. $O_2$ is produced by sunlight breaking apart water molecules, but once made, doesn't stick to solid ice very well and is thus exhaled into a ring atmosphere. There might ultimately be some implications for ring composition of such an increase in O and $O_2$ abundance.

## 10.5 RINGMOONS AND RINGLETS

As discussed in Chapter 9 and Section 10.3, moonlets are intimately intermingled in all planetary ring–moon systems. Those orbiting within and just outside Saturn's main rings are responsible for exciting a large number of spiral density waves, such as those in Figures 10.6 and 10.7, which may be used to determine their mass. Since we now have very good images of these moons giving their volume, we can easily determine their density. All the close-in ringmoons of Saturn have a density about half that of solid ice, indicating they are likely to be loose rubble piles rather than solid icebergs [37].

**The multi-stranded F ring**: Lying just outside the main rings, the F ring is where the direct interactions between moonlets and ringlets are most evident. "This is at the top of the list of things we didn't expect to see," said Voyager imaging team leader Brad Smith when the kinky, apparently braided or criss-crossing strands of Saturn's F ring were first imaged (Chapters 2 and 9; [38]). The F ring was originally thought to be a "shepherded" or confined ring, because of the two 100-km moonlets Voyager found to straddle it, a configuration that had been proposed to explain the narrow, but dense,

**Figure 10.26.** The F ring and its perhaps inappropriately-named "shepherd moons"; the larger (inner) moon Prometheus is shown here and the smaller (outer) moon Pandora is shown on the opposite page (see also Figure 10.28). Both are noticeably elongated, and may be just piles of loosely packed rubble. The five–seven strands seen here, covering perhaps 600 to 700 km, are more numerous and widely spaced than seen by Voyager.

rings of Uranus (Chapters 2 and 7). However, time and new discoveries have challenged this belief (Chapter 9; Section 10.1 [39]).

Cassini observations have already revealed that the entire F-ring region is indeed very dynamic and has changed since Voyager. For example, Voyager saw the F ring to be composed of two or three (and sometimes only one) narrow strands, plus a diffuse halo that extended out a few hundred kilometers and a faint belt of material that extended inwards, but not outwards, of the strands. Cassini sees at least five strands, with total radial range several times larger than Voyager, and a diffuse halo that extends to a full width of nearly 2,000 km (Figure 10.26).

**Figure 10.26** (*cont.*)

The strands seen by Cassini seem to be organized globally into a sort of spiral pattern—not a spiral density wave but merely a structure that results when a large clump of material is created instantaneously and stretched out into a spiral by its differential motion [40]. What might have caused this rather substantial event is not known, but one of the small moons discovered by Cassini in the F-ring region might be playing a role. The suspect object, originally called 2004S6, or S6 for short, was hardly visible at all when first spotted [41], but on subsequent sightings seems to have become a formidably bright clump with a noticeable trail of debris (Figure 10.27 [42]).

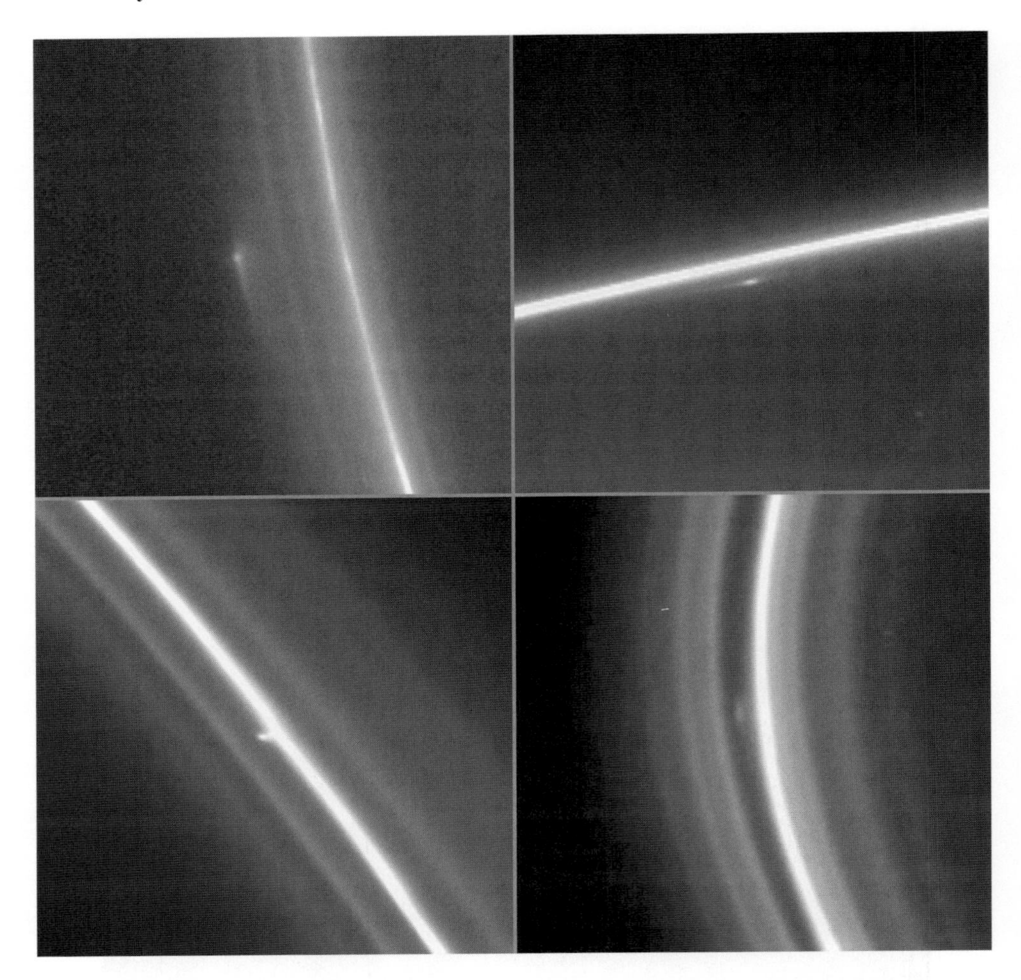

**Figure 10.27.** (Top) Two different images of what is thought to be the same object, called 2004S6, and (bottom) two other new objects with different orbits. Notice that the object is sometimes seen well inside the main strands of the F ring, and sometimes seen far outside. Its orbit appears to cross the F ring at a location where a disturbance began that might be responsible for a spiral nature of all the strands shown here. Notice the arc of debris associated with S6 in two images; it's not just a single object, and nothing of this sort was seen in the discovery image taken a few months earlier, in which it is just barely visible in similar observing geometry [41].

Moreover, its orbit seems to be highly eccentric, causing it to careen through the entire F-ring region from its inner to its outer perimeter—crossing through the strands in the process and, apparently, colliding with something. Perhaps this same collision not only disturbed the F-ring strands but also produced the debris trail of S6—or perhaps the debris trail was produced first in an unrelated collision with another ricocheting moonlet, and made it possible for S6 to have such a large effect on the F ring. With the occasional glitches seen in the orbits of Prometheus and Pandora being explained by orbital chaos [43], and with dynamical models suggesting even more chaos in the region between them, it's not surprising to see this kind of behavior. How many more of these moonlets are there? Is the entire F-ring region something like a mini-asteroid belt, with constant collisions generating clumps of material which are subsequently swept up by other moonlets? [44]. Its presence right at the edge of the Roche zone makes it possible that accretion might be proceeding, at least until one growing moonlet has an unfortunately violent encounter with another.

As time goes by, the inner ringmoon Prometheus, on an eccentric orbit, comes closer to passing right through the F-ring strands itself. This will happen in 2010, but even now it is getting close enough to cause significant changes in the orbits of F-region particles in its vicinity. For example, Figure 10.26 (left) shows an example of "drapes" or "gores" in the F-ring region (top of image) caused by massive Prometheus merely passing nearby. Dynamical models of this structure are in excellent agreement with the observations [45]. Figure 10.26 (left) also clearly shows Prometheus tearing a strand of material from the F-ring strands. This is not the behavior expected of a moon which is shepherding, or confining, a ring. Figure 10.26 (right) and Figure 10.28 show the outer "shepherd" Pandora. The clearly prolate (potato-like) shape of these ringmoons (Figures 10.26 and 10.28) suggests they are deforming into stable figures in

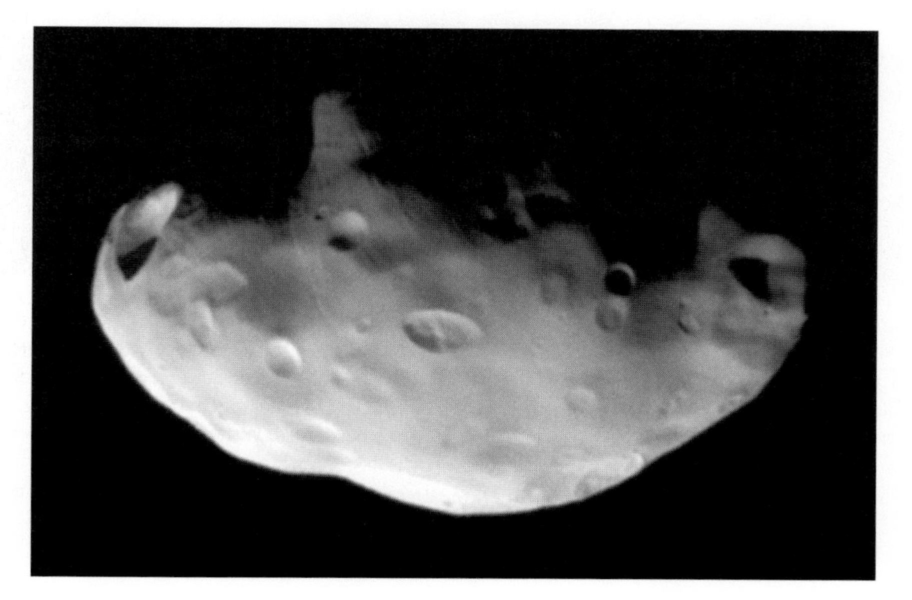

**Figure 10.28.** A Cassini closeup image of the ringmoon Pandora, which orbits outside the F ring. Pandora is nearly 100 km in long dimension, and has an internal density of about 0.5—thus is probably a loose rubble pile or agglomerate of large chunks with empty spaces between, covered by rubble. Its smooth surface implies recent resurfacing, probably by F-ring region material.

Saturn's strong self-gravity. They are sufficiently far away from Saturn to be able to grow by accreting smaller particles, so may have been destroyed and re-accreted several times in Saturn's history (see Section 10.7).

**Embedded moonlets** Late in the planning stages of the Voyager 2 encounter, it became a concern to some members of the imaging team that inadequate attention had been paid to looking for moonlets actually embedded in the rings. An attempt was made to address this, but was concentrated only on two gaps in the Cassini Division which, to this day, have shown no evidence for moonlets [46]. Not until three years after the Voyager encounters was the first key evidence found which led to the discovery of Pan in the Encke gap [47], and Pan itself was not found until 10 full years after the images in which it was discovered had been taken [48] (see Chapters 2 and 9). Cassini repeated that sequence of observations in its first few hours in orbit (Figure 10.29). Moreover, after a few months, Cassini discovered another embedded moonlet

**Figure 10.29.** Cassini image taken just after SOI, showing the Encke gap with its wavy edge and moonlet wake (left side) caused by a recent encounter with its embedded moonlet Pan, which orbits in the center of the gap (but out of the frame to the top). Several faint ringlets are seen inside the gap, each probably deriving from some other small, as-yet-undiscovered object. The middle ringlet is associated with Pan.

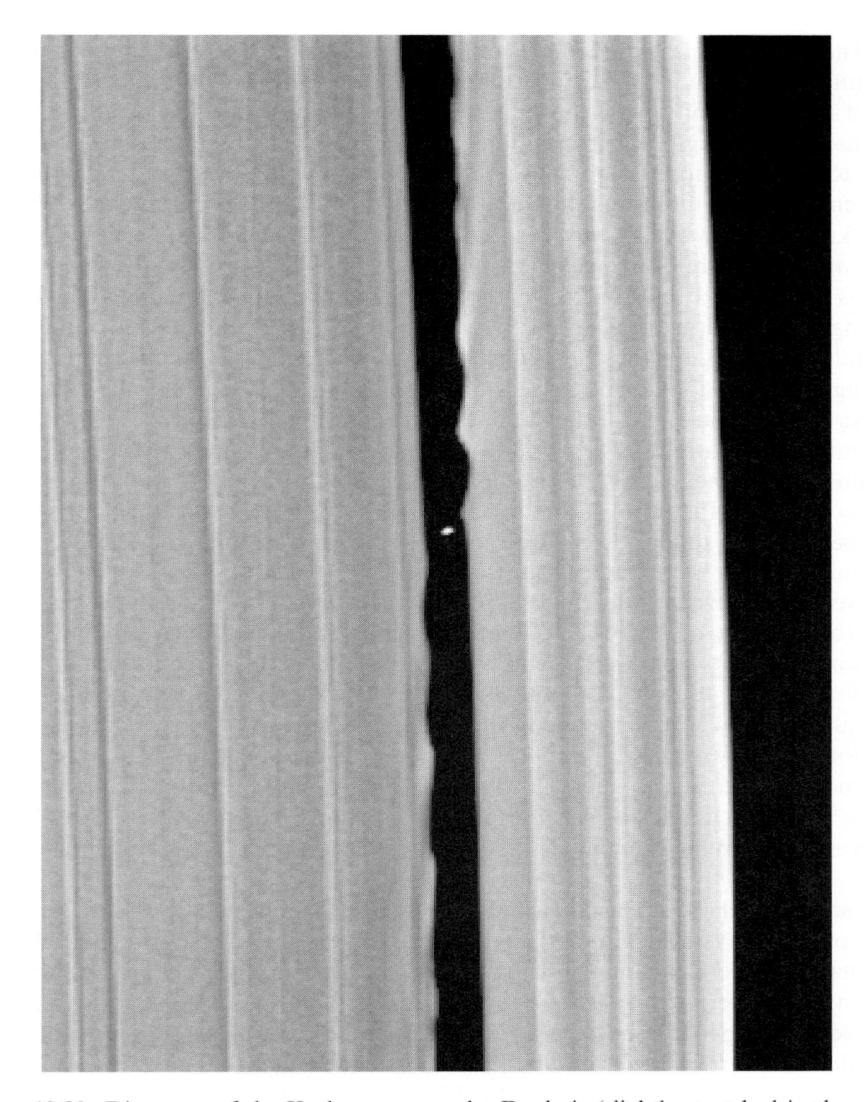

**Figure 10.30.** Discovery of the Keeler gap moonlet Daphnis (slightly stretched in the radial direction for visibility). The Encke gap (Figure 10.29, about 300 km across) is ten times the width of the Keeler gap, and its moonlet Pan is 30 times the mass of Daphnis. The edge waves of Daphnis are slightly peaky because it is on an eccentric orbit.

having similar effects on the edges of the Keeler gap, only 30 km across (Figure 10.30). The Keeler gap moonlet, called Daphnis, is on an eccentric orbit, which complicates the physics of its edge waves, but to a good approximation, its size and probable mass are in fairly good accord with the same theory that explains how the wider Encke gap is kept open by the larger moon Pan.

The prominent wavy inner edge in Figure 10.29 is ring material that has recently encountered Pan, and has been perturbed onto eccentric, nested orbits (Pan itself is out of the frame, off to the top by a few dozen frame widths). A small but steady increase in wavelength of perturbations that occur with increasing distance from Pan leads to the fan-shaped "moonlet wake" structure inward of the gap, appearing very different from the more tightly wrapped spiral density waves nearby [49]. The very same kind of edge wave structure is seen in Figure 10.30, and in fact here the moonlet Daphnis (artificially elongated) is visible in the gap. The direction of orbital motion here is from top to bottom, so material inwards of Daphnis, closer to Saturn, is moving faster and passes Daphnis by, and we find the inner edge wave to lead the moon. Conversely, for the outer edge, material where the waves are seen is actually lagging behind Daphnis and was just passed by it recently. Thus, the edge waves always lie "downstream" of the moonlet that causes them; the physics (Chapter 9) is very much like water in a stream flowing over a rock, making a ripple pattern that remains fixed to the rock even as water flows merrily through it. In movies made from images like these, the entire moon–wave pattern is seen moving together [50]—thus, at the speed of the moon. The Keeler gap edge waves are more "peaked" because Daphnis is on an eccentric orbit [51].

Notice the several faint ringlets in the Encke gap—one in the center, at or very near to the orbit of Pan. These contain clumps and kinks that come and go on timescales of days and weeks, and are surely related to bodies orbiting at these locations that are too small to have yet been seen. Several incomplete arc-like ringlets were also seen by Voyager—one in the center, as here, and one at a different radial location than any of these. Temporal changes are therefore common and frequent in the Encke gap, at least, where no doubt a small family of mountain-sized objects orbits, perhaps chaotically because of their proximity to each other. Searches are underway for similar phenomena in all of the other "empty" gaps in the main rings.

**Even smaller embedded chunks and shards** The question quickly arises as to what other objects may exist in the rings, with sizes between the largest "ring particle" and the smallest embedded moonlets capable of clearing a gap, such as Pan, Daphnis, and their as-yet-undetected cohorts. It was suggested in the Voyager era that such moonlets might perturb the material around them, accounting perhaps for some of the irregular structure in the B ring [52]. The perturbations that even smaller moonlets would exert on surrounding ring material were calculated more recently, and show small local disturbances called "propellors" [53]. These structures are, essentially, greatly shrunken versions of the edge wave pattern shown for the Keeler gap moonlet Daphnis in Figure 10.30; simply imagine the moonlet to shrink in size until the gap closes around it, leaving only local perturbations inside (leading) and outside (trailing) the object. The Cassini ISS team has recently discovered precisely such small structures (Figure 10.31, left) and thus inferentially the 100-m size objects responsible for them [54].

The Cassini discoveries were made in two separate images of the unlit face of the A ring in a single set of four of the highest resolution images taken by Cassini, during the SOI time period (Figure 10.2). That even this tiny sample of the rings revealed four such objects implies that there are many of them present, on statistical grounds, but unfortunately there are no further Cassini observations planned with resolution approaching that of these images (50 m/pixel). Figure 10.31 (right) indicates that the statistical abundance of these 100-m diameter objects is consistent with a continuous distribution ranging all the way from the largest ring particles (10-m diameter) to the known embedded moons Pan and Daphnis. The distribution is so steep that these shards do not represent a significant fraction of the total mass of the rings, which resides in the largest ring particles.

## 10.6 EVOLUTIONARY PROCESSES AND TIME VARIATIONS

Rings evolve structurally and compositionally. The evolution processes are both intrinsic (viscous spreading, gravitational wakes and possibly radial drift of large objects, and growth and disruption of large particles) and extrinsic (spiral wave torques at resonances with external moons, meteoroid bombardment, and loss of charged grains to the planet). These processes have been reviewed in Chapter 9 and elsewhere [55]. Here we merely sketch how Cassini (and very recent Earth-based) observations have changed our views or opened up new perspectives.

Torques between the rings and ringmoons were originally calculated to give a characteristic timescale of tens of million years [1] for reaching the current configuration if the ringmoons started at the edge of the rings. The newly calculated low density for these ringmoons increases these timescales by about a factor of 2. It was once thought that careful measurements of the orbit period of the ringmoons, between Voyager and Cassini, might allow this recession rate to be determined; however, evidence for chaotic interactions between the ringmoons seems to preclude this possibility [2]. Some dynamical models [56] raise the possibility that differential evolution leads to resonant interactions between ringmoons. The ensuing strong orbital perturbations lead to mutual collisions and disruption, after which re-accretion would have to occur before the rubble could once again act dynamically as a moon. Alternatively, resonances could tie these evolving ringmoons to more massive, outer moons, halting their evolution. These studies estimated a frustration of outward evolution of the inner ringmoons by some extra tens of millions of years, but this estimate is uncertain and seems unlikely to increase evolution lifetimes to the age of the solar system. Moreover, even if resonant trapping can halt the outward evolution of the ringmoons, there is nothing comparable to halt the associated inward evolution of the ring material, so the presence of the A ring, at least, where most of the density waves lie, remains a puzzle.

**Meteoroid bombardment** A constant hail of interplanetary meteoroids bombards the rings at velocities of tens of km/s, carrying more energy per unit mass than TNT. Ring particles are cratered or destroyed in these impacts (so we expect their composition to be well mixed), but the impact ejecta don't *escape* the rings, so material merely gets redistributed. This process is potentially important for both structural and compositional evolution of the rings. In terrestrial geology, slow erosion processes acting on

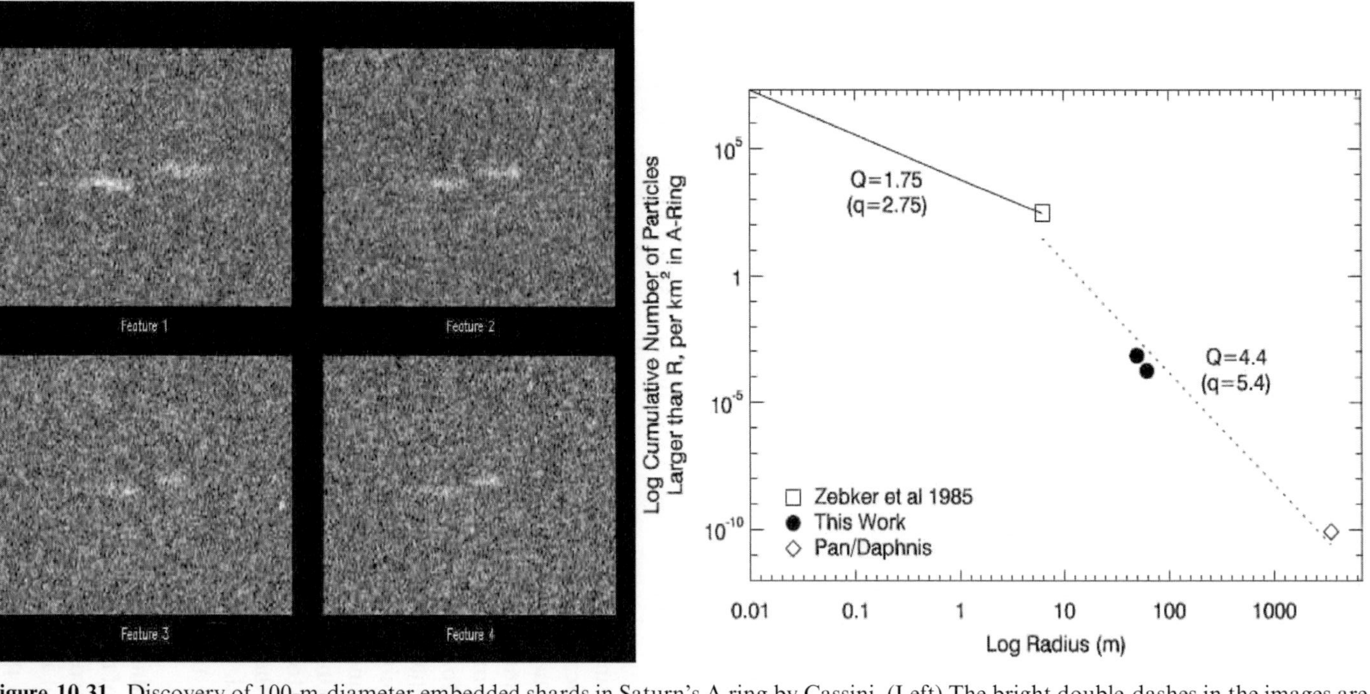


**Figure 10.31.** Discovery of 100-m-diameter embedded shards in Saturn's A ring by Cassini. (Left) The bright double-dashes in the images are regions of perturbed, perhaps slightly cleared or bunched up, particles. The perturbations which lead (trail) the moon are slightly inwards (outwards) of it as expected for relative orbital motion. Features such as these had been predicted theoretically prior to the Cassini encounter, and the mass of the object itself, lying unseen between the two dashes in each set, can be estimated from the length of the pattern. (Right) The abundance of shards such as these can be derived from the fractional area of the discovery frames. Their sizes lie near a power-law size distribution $n(r) = n_0(r/r_0)^{-q}$ between the largest ring particles and known moons in empty gaps.

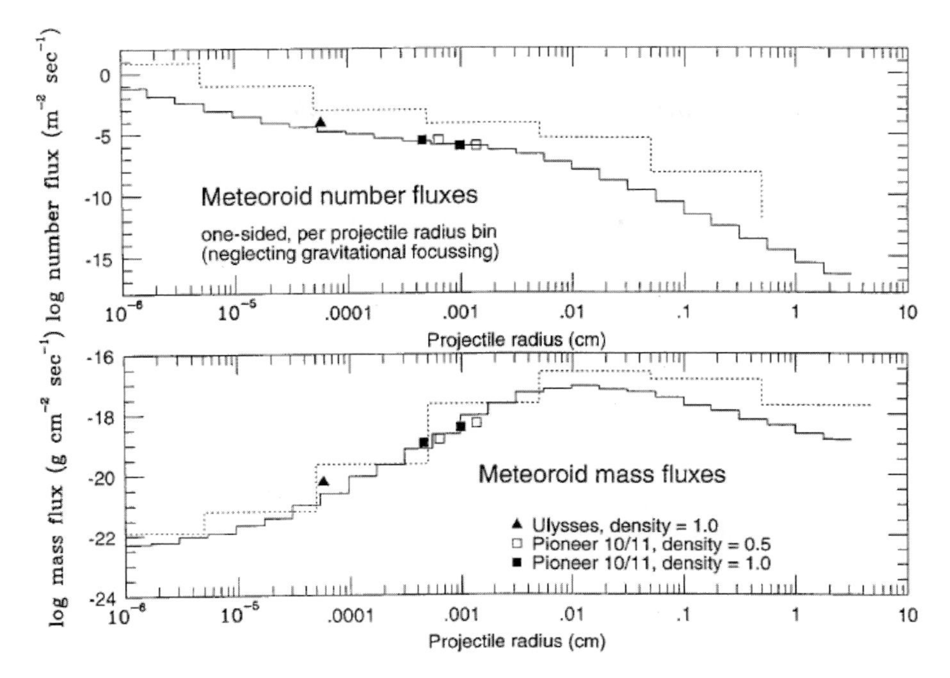


**Figure 10.32.** Possible size and mass distributions for interplanetary meteoroids bombarding Saturn's rings. The lines show the distribution of meteoroids at Earth; the points are directly measured by spacecraft in the 10-AU region. The shape of the curve at 10 AU is, however, unknown.

tiny length-scales can remove mountains, given enough time. The current observations of the meteoroid flux in the environment of Saturn, by Pioneer 10 and 11, and by Ulysses [57], lie very close to the mass distribution measured at 1 AU, which covers a wider mass range (Figure 10.32 [58]).

If the same form of the mass distribution prevails at Saturn (which is uncertain, and the goal of the observations described below), the huge collecting area of the rings means that they absorb their own mass over the age of the solar system. If this is true, it provides a second "young ring" constraint as described in Chapter 9, and below.

Of the several uncertainties associated with this process, the inbound mass flux is certainly the most profound. Note in Figure 10.32 that the observations don't sample the likely peak of the mass flux. Because of the pointing requirements of Cassini's outbound cruise to Saturn, its CDA instrument wasn't able to improve on past measurements of the interplanetary mass flux [59]. There are several other possible ways to measure this flux, albeit somewhat less directly. All involve observing the consequences of impacts by different sized particles in different ways, and thus require some process modeling to interpret the observations.

*Saturn's icy moons and the F ring*: The airless moons are themselves excellent meteoroid detectors; properly designed close flybys of these moons to measure dust grains ejected from their surfaces might still provide a good estimate of the inbound flux. These observations will need to be made late in Cassini's nominal mission, or in its extended mission [60]. An uncertainty here is the mass of dust particles of measurable sizes that are ejected by a given projectile mass, and their ejection velocities (which determine the fraction that escapes the surface of a moon of arbitrary size); these will depend on the nature of the moon's surface. Also, some Voyager observations suggested that the optically thin F ring itself might be a sufficiently large target that large bursts of particles might be seen, caused by impacts of 10-cm meteoroids [61]. A number of remote-monitoring observations have been taken to address this goal, but more time and analysis is needed for their study.

*Spokes*: Impacts on the rings have been advocated as a trigger for "spokes" (Chapter 9). As described in Chapter 9, spoke observations have diminished as the ring opening angle to the Earth and Sun has increased. Both Voyager 1 and 2 encounters, when spokes were prominent, occurred when the rings had a small opening angle. Hubble Space Telescope observations over nearly a decade have clearly shown that the abundance of spokes drops sharply as the ring becomes more open as seen from the Sun and Earth [62]. Cassini approach observations followed this same trend. One suggestion was that a vertical distribution of spoke particles which is thicker than the main rings—maybe only by several times—could explain the observations [52]. However, this hypothesis predicted the reappearance of spokes when Cassini first observed the rings from a small elevation angle in October 2004, and this did not occur.

Another suggestion [63] involves the seasonal charging of the rings by sunlight, perhaps (or even probably) in addition to a vertically extended layer. When sunlight hits the rings, electrons are boiled off their surface into a surrounding layer. Thus, the main rings are positively charged and a net electric field exists between them and the surrounding electron cloud. If tiny dust grains are ejected above the rings, perhaps to form a spoke, they become charged by the electrons and pulled back down to the positively charged ring by the net field. This electrostatic vacuum may prevent spokes from becoming visible until the ring opening angle to the Sun is small, weakening this photo-charging process.

Cassini has now seen a few faint spokes, at high phase angle on the unlit face of the ring (Figure 10.33). Shortly after this observation in September 2005, the spacecraft began a long series of equatorial orbits from which it had just emerged as this chapter was written, and a few more spokes have been seen. As time goes on, the solar elevation angle will decrease further, and we will probably see more spoke activity. Surely by the end of the mission, when the opening angle to the Sun will be similar to Voyager geometry, spokes will again be frequent and prominent.

By counting spokes (once we start observing them again, and if we can convince ourselves that they are related to impacts) we might be able to infer the impact rate of meteoroids of some size—but it is not known what size projectile creates a spoke of a given size. The currently most popular theory [64] does give such an estimate (roughly meter-size), but this theory has been challenged recently [65], and thus the whole process connecting an impact with an observed spoke remains controversial. An alternative, but less widely read, model has been proposed for spoke formation

**Figure 10.33.** Image of a few narrow spokes on the unlit face of Saturn's outer B ring (the dark gap is the Huygens gap in the Cassini Division), seen in highly forward-scattered light which makes them especially easy to observe. These narrow and fresh-looking spokes were actually seen going into the shadow of the planet on the rings. This indicates that spokes do not have to form inside the shadow but remains consistent with impact trigger theories.

[66], which can also be triggered by meteoroid impacts. Certainly, more theoretical work needs to be done to resolve these issues before we can even be convinced that spokes are caused by meteoroid impacts, or use their properties to derive an impact rate.

*Impact flashes*: Cassini had also planned a number of observations of a different sort, related to impacts, that would not depend on seeing the shadowy, flickering spoke features themselves, but instead detecting the impacts directly. The UVIS instrument is designed for rapid sampling of very hot, ultraviolet light-emitting stars as they are occulted by the rings and planet. It was planned to use this same capability, staring for hours into Saturn's shadow on the rings, where impacts are expected to be the most energetic, to detect flashes of light coming from the hot gas associated with these rather violent explosions. For example, flashes of light have been observed from impacts of Leonid meteors on the Moon [67]. Although over 100 hours of observing time were devoted to these observations, nothing has been seen resembling an impact flash. Furthermore, new modeling of the impact process has found that the hot gas capable of emitting ultraviolet photons that could be seen by UVIS is blanketed in cooler, opaque gas which emits light only at visible wavelengths where UVIS is blind—explaining the lack of detections [68].

*Plasma waves*: A unique and intriguing observation of a new phenomenon was made by the RPWS instrument during the SOI time period, as Cassini flew immediately above the rings. During the 30 minutes or so it took Cassini to traverse the rings, dozens of nearly pure "tones" of radio waves with different, but well-defined, frequencies were heard, each for only a few seconds. It has been suggested that these tones are a form of plasma wave, excited by the hot, expanding, ionized plasma of the impact event (remember this hot gas has a temperature of 10,000 K), and traveling along well-defined paths that vary with frequency [69]. Counting these events and estimating the areal coverage of the spacecraft's sensitivity, one can derive an impact rate. What is missing is a theory which connects the size of the projectile to the strength of the wave; without such a connection we can't say what meteoroid size is associated with the observed event rate. This is another excellent problem for the future.

Ultimately, making use of one, or hopefully several, of these necessarily indirect approaches, we hope to obtain a firmer grasp on the mass flux of meteoritic material bombarding the rings.

*Effects of meteoroid bombardment*: Meteoroid bombardment has both structural and compositional implications, as discussed in Chapter 9. The primary effect of relevance to Cassini observations thus far is the evolution of ring composition. It was noted [70] that, because meteoroids at Saturn are likely to be rich in carbon and silicate material, only a small fraction of a ring mass, if mixed uniformly into the ring material, could make the particles much darker than they actually are: cosmic "pollution". Indeed the ring particles at Jupiter, Uranus, and Neptune do appear to be quite dark (see Chapters 6, 7, and 8). This was the second indication of a short ring timescale—or young age—and is independent of the density wave arguments mentioned in Section 10.3. Subsequent detailed modeling of how ejecta are absorbed and redistributed within the rings [10] showed that certain characteristic radial profiles of ring composition would result—in particular, the optically thinner, less massive rings such as the C ring and Cassini Division should be more polluted, more closely resembling dark, neutral-colored cometary material (as is the case), and ring composition should vary smoothly across even abrupt ring optical depth discontinuities. These models focused on the characteristic inner edge of the B ring, which is very similar in appearance to the inner edge of the A ring (both structures are also candidates for being maintained by redistribution of meteoroid ejecta [71]). Initial comparisons were with radial profiles of ring color, which indeed seemed to vary smoothly as the models predicted.

Cassini observations by VIMS seem to be supporting this effect. VIMS observes not only a color, but compositional variation associated with specific spectral features. Indeed, the C ring and Cassini Division do seem to contain a larger fraction of iron-rich, plausibly silicate, material, and the optically thin F ring shows evidence for C–N-based organics associated with dark material on Phoebe (a primitive, comet-nucleus-like object) [72] and thus also plausibly with interplanetary debris. In Figure 10.17 we showed some radial profiles of the strength of water-ice bands as measured by VIMS.

The smooth variation of these profiles agrees well with the smooth variation of color seen previously. Other VIMS metrics—such as the strength of the iron-(silicate?) feature—also vary smoothly across these abrupt boundaries in a similar way. There are a number of parameters in these models—such as the mass-to-optical-depth ratio as a function of radius—of which Cassini observations of different types can ultimately improve our understanding.

**Yet another short timescale** It has been suggested that tiny grains, once charged by a number of processes—perhaps impact events, perhaps photosputtering—can quickly spin down magnetic field lines into the planet's upper atmosphere and be lost [73]. This loss process primarily affects the C ring and inner B ring, rather than the A ring as do density wave torques. Like all previous models, it has uncertain parameters, but the best estimates indicate that the entire C ring can be lost this way in some few tens of million years.

## 10.7 RING ORIGIN

The origin of Saturn's rings remains an unsolved problem. It's not so hard to imagine how the ring systems of the other three gas giants formed—maybe are re-forming even today—because none of them represents much mass, so could be created by easy disruption of small moonlets. Moreover, all of their particles are thoroughly as dark as anything else in the outer solar system, so there is no need to create them recently. Saturn's main rings are fundamentally unlike the other three known ring–moon systems in their mass and generally unpolluted icy brilliance. Because of the difficulty of retaining their current configuration for the age of the solar system in the face of several proposed removal and evolutionary processes, the question of origin takes on a modern, rather than a primordial, context. This tends to raise eyebrows and questions, because by several hundred million years ago, the solar system is thought to have left behind the cataclysmic, violent, collisional formation environment of its youth behind and settled down into a calm maturity. The rings have the mass of Mimas, a moon with radius of about 200 km. To break something that size down into rubble, of which Pan and Daphnis may represent the largest fragments, takes an unusually energetic event, and doing it recently is even more difficult because of the rarity of such events. We believe that the large craters seen on heavily cratered solid surfaces in the solar system are ancient, representing a final sweepup of the planetesimals, embryos, and minor planets that created the Earth and its family. We expect very few of such large impacts to have happened in "modern" times of "only" 400–500 million years ago—the Age of Fishes here on Earth.

There are two fundamentally different parentages one can imagine for the rings, presuming them to be of secondary origin (the debris of a destroyed parent, rather than direct and primordial nebula condensates). Some locally formed moon like Mimas, for example, might have been destroyed by a collision with some incoming projectile [74]. Alternatively, some interloper from beyond the Saturn system might have grazed by Saturn too closely, and been torn apart by its tides [75]. In either case the parent would need to be fairly large—probably larger than the amount of mass currently found in the rings—both to allow for capture inefficiency, and to allow it to have differentiated prior to being destroyed.

More specifically, Saturn's irregular moon Phoebe is very dark, and looks not at all like the rings. It probably represents a truly primitive object—containing a mix of primordial water ice, silicates, and carbon-rich tarry organics in roughly equal proportions [76]. Such a primitive object would need to melt, to separate its water ice (along with perhaps some light, reddish organics) into an icy outer shell, with its dark silicate and organic material settling into a denser core perhaps. Saturn's regular icy moons generally look this way. Even Iapetus, under its one soot-covered face, seems to be primarily a bright, white, object with a density very close to that of pure water. As another example, Neptune's major moon Triton is thought to be a captured Kuiper Belt object, and indeed has an extremely bright, icy surface—but this could perhaps have been produced by heating associated with its capture. However, in recent years, several large, Sun-orbiting Kuiper Belt objects have been discovered which are much brighter than the normally assumed 4% reflectivity [77]. The fact that most of Saturn's icy moons, like the rings, are already very water-rich, for reasons also not well understood, might seem to favor an intrinsic origin. However, a difference that has been noticed previously but becomes more obvious with Cassini observations (e.g., Figure 10.20) is that the rings are not the same color as the icy moons of Saturn; they are much redder. Is this a hint of a different formation environment? Triton is quite pink—its visual spectrum in fact is fairly similar to that of the rings. With additional Cassini observations of the composition of the rings, and of Saturn's moons, these differences will be characterized more carefully. For fun, we note one speculative possibility raised by the Cassini observations of abundant O and $O_2$ in the ring atmosphere: Is $O_2$ recycled again and again in the rings until it finally encounters one of the few iron atoms there, and is the ring redness simply rust?

## 10.8 SUMMARY

As of this writing, Cassini observations of Saturn's rings are less than one-third complete. Seven optimized orbits have been devoted partly to ring studies, and approximately thirty-eight more orbits remain in the tour, of which rings observations represent a decent fraction. Furthermore, if all goes well, Cassini is planning an extended mission which will continue until after the Sun crosses the edge-on rings and the north face of the rings is once again lit. Even a moment's thought reveals the depth of return which will benefit future ring scientists; the two Voyager encounters contained perhaps a few weeks worth of data each, of which the day or two near closest approach of each spacecraft represented the richest part. *Each* periapse pass of Cassini contains as much data as one Voyager encounter, taken with more and better instruments. Voyager had a single radio occultation (Cassini will have 14) and a single stellar occultation (Cassini will have about 80).

Analysis of the two Voyager encounters required nearly 20 years to complete, and the Cassini data set is nearly 100 times as large and complex. This is partly because

Cassini has entirely new instruments and partly because of the extended and complex orbital tour. The ring scientists responsible for planning these observations are a small and dedicated crew, but the necessity for assuring that the proper observations get designed, taken, and calibrated has taken priority over the fun of analyzing and publishing results once they arrive—a situation which continues as this chapter is written and will continue for several more years, because ongoing updates to the mission profile require ongoing changes in observing plans. It's reminiscent of a line from a popular song relating to poker: "Never count your money while you're sitting at the table—there'll be time enough for counting when the dealing's done." Generations of ring scientists will be playing with the cards dealt by this remarkable mission, still in its early days.

## 10.9 NOTES AND REFERENCES


[1] See, for example, Goldreich, P., Tremaine, S., 1982, "The dynamics of planetary rings", *Annual Reviews of Astronomy and Astrophysics* **20**, 249–283.

[2] French, R. C., McGhee, C. A., Dones, L., Lissauer, J. J., 2003, "Saturn's wayward shepherds: The peregrinations of Prometheus and Pandora", *Icarus* **162**, 143–170; Goldreich, P., Rappaport, N., 2003, "Origin of chaos in the Prometheus–Pandora system", *Icarus* **166**, 320–327.

[3] Cuzzi, J. N., Burns, J. A., ,1988, "Charged particle depletion surrounding Saturn's F ring—Evidence for a moonlet belt?", *Icarus* **74**, 284–324; Barbara, J., Esposito, L. W., 2002, "Moonlet collisions and the effects of tidally modified accretion in Saturn's F ring", *Icarus* **160**, 161–171; Poulet, F., Sicardy, B., 2001, "Dynamical evolution of the Prometheus–Pandora system", *Monthly Notices of the Royal Astronomical Society* **322**, 343–355.

[4] Showalter, M. R., 2004, "Disentangling Saturn's F ring. I. Clump orbits and lifetimes", *Icarus* **171**, 356–371.

[5] McGhee, C. A., Nicholson, P. D., French, R. G., Hall, K. J., 2001, "HST observations of Saturnian satellites during the 1995 ring plane crossings", *Icarus* **152**, 282–315.

[6] Reitsema, H. J., Beebe, R. F., Smith, B. A., 1976, "Azimuthal brightness variations in Saturn's rings", *Astronomical Journal* **81**, 209–215; Marouf, E. A., Tyler, G. L., Zebker, H. A., Simpson, R. A., Eshleman, V. R., 1983, "Particle size distributions in Saturn's rings from Voyager 1 radio occultation", *Icarus* **54**, 189–211; Salo, H., 1992, "Gravitational wakes in Saturn's rings", *Nature* **359**, 619–621; Karjalainen, R., Salo, H., 2004, "Gravitational accretion of particles in Saturn's rings", *Icarus* **172**, 328–348.

[7] Because ring particles closer to the planet experience a stronger gravitational force than their neighboring particles just slightly farther from Saturn, they orbit at higher speeds. This steady variation of orbit velocity with distance from the planet is called differential rotation, and it leads to collisions which stir the particles.

[8] Esposito, L. W., Cuzzi, J. N., Holberg, J. B., Marouf, E. A., Tyler, G. L., Porco, C. C., 1984, "Saturn's rings: Structure, dynamics, and particle properties", in *Saturn*, edited by Gehrels, T. and Matthews, M. S.; Poulet, F., Cruikshank, D. P., Cuzzi, J. N., Roush, T. L., French, R. G., 2003, "Composition of Saturn's rings A, B, and C from high-resolution near-infrared spectroscopic observations", *Astronomy and Astrophysics* **412**, 305–316.

[9] Cooke, M., 1991, "Photometry of Saturn's C ring", PhD Thesis, Cornell University; Doyle, L. R., Dones, L. C., Cuzzi, J. N., 1989, "Radiative transfer modeling of Saturn's

outer B ring", *Icarus* **80**, 104–135; Dones, L., Cuzzi, J. N., Showalter, M. R., 1993, "Voyager photometry of Saturn's A ring", *Icarus* **105**, 184–215; Estrada, P., Cuzzi, J. N., 1996, "Voyager observations of the color of Saturn's rings", *Icarus* **122**, 251–272; Estrada, P. R., Cuzzi, J. N., Showalter, M. R., 2003, "Voyager color photometry of Saturn's main rings: A correction", *Icarus* **166**, 212–222.

[10] Cuzzi, J. N., Estrada, P. R., 1998, "Compositional evolution of Saturn's rings due to meteoroid bombardment", *Icarus* **132**,1–35.

[11] Meyer-Vernet, N., Lecachaux, A., Pedersen, B. M., 1998, "Constraints on Saturn's G ring from the Voyager 2 radio astronomy instrument", *Icarus* **132**, 311–320, and references therein.

[12] Porco, C. C., West, R. A., McEwen, A., Del Genio, A. D., Ingersoll, A. P., Thomas, P., Squyres, S., Dones, L., Murray, C. D., Johnson, T. V., Burns, J. A., Brahic, A., Neukum, G., Veverka, J., Barbara, J. M., Denk, T., Evans, M., Ferrier, J. J., Geissler, P., Helfenstein, P., Roatsch, T., Throop, H., Tiscareno, M., Vasavadal, A. R., 2003, "Cassini imaging of Jupiter's atmosphere, satellites, and rings", *Science* **299**, 1541–1547.

[13] See *Science* **307**, no. 5713, February 25, 2005, special issue on Cassini Saturn encounter.

[14] *http://saturn.jpl.nasa.gov/operations/saturn-tour.cfm*

[15] Lane, A. L., Hord, C. W., West, R. A., Esposito, L. W., Coffeen, D. L., Sato, M., Simmons, K. E., Pomphrey, R. B., Morris, R. B., 1982, "Photopolarimetry from Voyager 2: Preliminary results on Saturn, Titan, and the rings", *Science* **215**, 537–543; Holberg, J. B., Forrester, W. T., Lissauer, J. J., 1982, "Identification of resonance features within the rings of Saturn", *Nature* **297**, 115–120.

[16] Shu, F. H., 1984, "Spiral waves", in *Planetary Rings*, edited by Greenberg, R. and Brahic, A.; Greenberg, R., 1984, "Resonances", in *Planetary Rings*, edited by Greenberg, R. and Brahic, A.

[17] Colwell, J. E., Esposito, L. W., Stewart, G. R., 2006; "Density waves observed by Cassini stellar occultations as probes of Saturn's rings"; *37th Lunar and Planetary Science Conference*, March 13–17, League City, Texas, Abstract #1221; Marouf, E., French, R., Rappaport, N, Kliore, A., Flasar, M., Nagy, A., McGhee, C., Schinder, P., Anabtawi, A., Asmar, S., Barbinis, E, Fleischman, D., Goltz, G., Johnston, D., Rochblatt, D., Thomson, F., Wong, K., 2005, "Structure of Saturn's rings from Cassini diametric radio occultations", *Bulletin of the American Astronomical Society* **37**, DPS meeting #37, abstract #62.02.

[18] Horn, L. J., Cuzzi, J. N., 1996, "Characteristic wavelengths of irregular structure in Saturn's B and A rings", *Icarus* **119**, 285–310.

[19] Salo, H., Schmidt, J., Spahn, F., 2001, "Viscous overstability in Saturn's B ring. I. Direct simulations and measurement of transport coefficients", *Icarus* **153**, 295–315; Schmidt, J., Salo, H., Spahn, F., Petzschmann, O., 2001, "Viscous overstability in Saturn's B-ring. II. Hydrodynamic theory and comparison to simulations", *Icarus* **153**, 316–331.

[20] Rosen, P. A., Tyler, G. L., Marouf, E., Lissauer, J. J., 1991, "Resonance structures in Saturn's rings probed by radio occultation. II—Results and interpretation", *Icarus* **93**, 25–44.

[21] See Cuzzi, J. N., Lissauer, J. J., Esposito, L. W., Holberg, J. B., Marouf, E. A., Tyler, G. L., Boischot, A., 1984, "Saturn's rings: Properties and processes", in *Planetary Rings*, edited by Greenberg, R. and Brahic, A., pp. 73–199, for global profiles of ring optical depths.

[22] *Ibid.*; Nicholson, P. D., Hedman, M. M., Wallis, B., 2005, "Cassini-VIMS observations of stellar occultations by Saturn's rings", *Bulletin of the American Astronomical Society* **37**, DPS meeting #37, abstract #62.05; Ferrari, C., Spilker, L., Brooks, S., Edgington, S. G., Wallis, B., Pearl, J., Leirat, C., Flasar, M., 2005, "Azimuthal temperature variations in

Saturn's rings as seen by the CIRS spectrometer onboard Cassini", *Bulletin of the American Astronomical Society* **37**, DPS meeting #37, abstract #62.07.

[23] Mischchenko, M., Dlugatch, Zh. M., 1992, "Can weak localization of photons explain the opposition effect of Saturn's rings?" *Monthly Notices of the Royal Astronomical Society* **254**, 15P–18P; Hapke, B. W., Nelson, R. M., Brown, R. H., Spilker, L. J., Smythe, W. D., Kamp, L., Boryta, M., Leader, F., Matson, D. L., Edgington, S., Nicholson, P. D., Filaccione, G., Clark, R. N., Bibring, J.-P., Baines, K. H., Buratti, B., Bellucci, G., Capaccioni, F., Cerroni, P., Combes, M., Coradini, A., Cruikshank, D. P., Drossart, P., Formisano, V., Jaumann, R., Langevin, Y., McCord, T. B., Mennella, V., Sicardy, B., Sotin, C., 2006, "Cassini observations of the opposition effect of Saturn's rings. 2. Interpretation: Plaster of Paris as an analog of ring particles", *37th Lunar and Planetary Science Conference*, March 13–17, League City, Texas, Abstract #1466.

[24] Nelson, R. M., Hapke, B. W., Brown, R. H., Spilker, L. J., Smythe, W. D., Kamp, L., Boryta, M., Leader, F., Matson, D. L., Edgington, S., Nicholson, P. D., Filaccione, G., Clark, R. N., Bibring, J.-P., Baines, K. H., Buratti, B., Bellucci, G., Capaccioni, F., Cerroni, P., Combes, M., Coradini, A., Cruikshank, D. P., Drossart, P., Formisano, V., Jaumann, R., Langevin, Y., McCord, T. B., Mennella, V., Sicardy, B., Sotin, C., 2006, "Cassini observations of the opposition effect of Saturn's rings. 1", *37th Lunar and Planetary Science Conference*, March 13–17, League City, Texas, Abstract #1461.

[25] Esposito, L. W., Colwell, J. E., Larsen, K., McClintock, W. E., Stewart, A. I. F., Hallett, J. T., Shemansky, D. E., Ajello, J. M., Hansen, C. J., Hendrix, A. R., West, R. A., Keller, H. U., Korth, A., Pryor, W. R., Reulke, R., Yung, Y. L., 2005, "Ultraviolet imaging spectroscopy shows an active Saturnian system", *Science* **307**, 1251–1255.

[26] *http://photojournal.jpl.nasa.gov/catalog/PIA07762*; Porco, C. C., Helfenstein, P., Thomas, P. C., Ingersoll, A. P., Wisdom, J., West, R., Neukum, G., Denk, T., Wagner, R., Roatsch, T., Kieffer, S., Turtle, E., McEwen, A., Johnson, T. V., Rathbun, J., Veverka, J., Wilson, D., Perry, J., Spitale, J., Brahic, A., Burns, J. A., Del Genio, A. D., Dones, L., Murray, C. D., Squyres, S., 2006, "Cassini observes the active south pole of Enceladus", *Science* **311**, 1393–1401; Spahn, F., Schmidt, J., Albers, N., Hörning, M., Makuch, M., Seiß, M., Kempf, S., Srama, R., Dikarev, V., Helfert, S., Moragas-Klostermeyer, G., Krivov, A. V., Sremcêević, M., Tussolino, A. J., Economou, T., Grün, E., 2006, "Cassini dust measurements at Enceladus and implications for the origin of the E ring", *Science* **311**, 1416–1418.

[27] Pang, K. D., Voge, C. C., Rhoads, J. W., Ajello, J. M., 1984, "The E-ring of Saturn and satellite Enceladus", *Journal of Geophysical Research* **89**, 9459–9470; Terrile, R. J., Cook, A. F. III, 1981, "Enceladus: Evolution and possible relationship to Saturn's E-ring", Abstract #428, *Lunar and Planetary Science Conference* **XII**, p. 10.

[28] Roddier, C., Roddier, F., Graves, J. E., Northcott, M. J., 1998, "Discovery of an arc of particles near Enceladus' orbit: A possible key to the origin of the E ring", *Icarus* **136**, 50–59.

[29] Showalter, M. R., Cuzzi, J. N., 1993, "Seeing ghosts—Photometry of Saturn's G ring", *Icarus* **103**, 124–143; Throop, H., Esposito, L. W., 1998, "G ring particle sizes derived from ring plane crossing observations", *Icarus* **131**, 152–166.

[30] Nicholson, P. D., Clark, R. N., Cruikshank, D. P., Showalter, M. R., Sicardy, B., Cassini VIMS Team, 2006, "Observations of Saturn's rings by Cassini VIMS: Saturn orbit insertion data", *Icarus*, submitted.

[31] Clark, R. N., Brown, R., Baines, K., Bellucci, G., Bibring, J.-P., Buratti, B., Capaccioni, F., Cerroni, P., Combes, M., Coradini, A., Cruikshank, D., Drossart, P., Filacchione, G., Formisano, V., Jaumann, R., Langevin, Y., Matson, D., McCord, T., Mennella, V., Nelson, R., Nicholson, P., Sicardy, B., Sotin, C., Curchin, J., Hoefen, T., 2005, "Cassini

VIMS compositional mapping of surfaces in the Saturn system and the role of water, cyanide compounds and carbon dioxide", *Bulletin of the American Astronomical Society* **37**, DPS meeting #37, abstract #39.05.

[32] Cuzzi, J. N., Estrada, P. R., 1998, "Compositional evolution of Saturn's rings due to meteoroid bombardment", *Icarus* **132**, 1–35; Cuzzi, J. N., French, R. G., Dones, L., 2002, "HST multicolor (255–1042 nm) photometry of Saturn's main rings. I: Radial profiles, phase and tilt angle variations, and regional spectra", *Icarus* **158**, 199–223.

[33] Smith, B. A., Soderblom, L. A., Banfield, D., Barnet, C., Basilevsky, A. T., Beebe, R. F., Bollinger, K., Boyce, J. M., Brahic, A., Briggs, G. A., Brown, R. H., Chyba, C., Collins, S. A., Colvin, T., Cook, A. F. H., Crisp, D., Croft, S. K., Cruikshank, D., Cuzzi, J. N., Danielson, G. E., Davies, M. E., De Jong, E., Dones, L., Godfrey, D., Goguen, J., Grenier, I., Haemmerle, V. R., Hammel, H., Hansen, C. J., Helfenstein, C. P., Howell, C., Hunt, G. E., Ingersoll, A. P., Johnson, T. V., Kargel, J., Kirk, R., Kuehn, D. I., Limaye, S., Masursky, H., McEwen, A., Morrison, D., Owen, T., Owen, W., Pollack, J. B., Porco, C. C., Rages, K., Showalter, M., Sicardy, B., Simonelli, D., Spencer, J., Sromovsky, B., Stoker, C., Strom, R. G., Suomi, V. E., Synott, S. P., Terrile, R. J., Thomas, P., Thompson, W. R., Verbiscer, A., Veverka, J., 1989, "Voyager 2 at Neptune: Imaging science results", *Science* **246**, 1422–1449.

[34] Wallis, B. D., Spilker, L. J., Pilorz, S. H., Pearl, J. C., Altobelli, N., Edgington, S. G., Flasar, F. M., Cassini CIRS Team, 2005, "CIRS observations of a thermal enhancement near zero phase in Saturn's rings", AGU Fall Meeting 2005, abstract #P33B-0250; Spilker, L. J., Pilorz, S. H., Ferrari, C., Leyrat, C., Wallis, B. D., Brooks, S. M., Edgington, S. G., Altobelli, N., Flasar, F. M., Pearl, J. C., Showalter, M. R., Achterberg, R. K., Nixon, C. A., Romani, P. N., Cassini CIRS Team, 2006, "Cassini CIRS observations of thermal differences in Saturn's main rings with increasing phase angle", *37th Lunar and Planetary Science Conference*, March 13–17, League City, Texas, Abstract #2299.

[35] Spilker, L. J., Pilorz, S. H., Wallis, B. D., Brooks, S. M., Edgington, S. G., Flasar, F. M., Pearl, J. C., Showalter, M. R., Ferrari, C., Achterberg, R. K., Nixon, C. A., Romani, P. N., Cassini CIRS Team, 2005, "Cassini CIRS observations of Saturn's rings", *36th Lunar and Planetary Science Conference*, March 14–18, League City, Texas, Abstract #1912.

[36] Waite, J. H. Jr., Cravens, T. E., Ip, W.-H., Kasprzak, W. T., Luhmann, J. G., McNutt, R. L., Niemann, H. B., Yelle, R. V., Mueller-Wodarg, I., Ledvina, S. A., Scherer, S., 2005, "Oxygen ions observed near Saturn's A ring", *Science* **307**, 1260–1262.

[37] Rosen *et al.* [20]; Nicholson, P. D., Hamilton, D. P., Matthews, K., Yoder, C. F., 1992, "New observations of Saturn's coorbital satellites", *Icarus* **100**, 464–484.

[38] Smith didn't elaborate as to whether the list had been prepared prior to encounter or as new discoveries mounted.

[39] Showalter, M. R., Burns, J. A., 1982, "A numerical study of Saturn's F-ring", *Icarus* **52**, 526–544; Cuzzi, J. N., Burns, J. A., 1988, "Charged particle depletion surrounding Saturn's F ring: Evidence for a moonlet belt", *Icarus* **74**, 284–324; Scargle, J., Cuzzi, J., Dobrovolskis, A., Dones, L., Hogan, R., Levit, C., Showalter, M., Young, K., 1993, "Dynamical evolution of Saturn's rings", *Bulletin of the American Astronomical Society* **25**, 1103 (abstract only); see also reference [2] above.

[40] Charnoz, S., Porco, C. C., Déau, E., Brahic, A., Spitale, J. N., Bacques, G., Baillie, K., 2005, "Cassini discovers a kinematic spiral ring around Saturn", *Science* **310**, 1300–1304.

[41] Porco, C. C., Baker, E., Barbara, J., Beurle, K., Brahic, A., Burns, J. A., Charnoz, S., Cooper, N., Dawson, D. D., Del Genio, A. D., Denk, T., Dones, L., Dyudina, U., Evans, M. W., Giese, B., Grazier, K., Helfenstein, P., Ingersoll, A. P., Jacobson, R. A., Johnson, T. V., McEwen, A., Murray, C. D., Neukum, G., Owen, W. M., Perry, J., Roatsch, T.,

Spitale, J., Squyres, S., Thomas, P., Tiscareno, M., Turtle, E., Vasavada, A. R., Veverka, J., Wagner, R., West, R., 2005, "Cassini imaging science: Initial results on Saturn's rings and small satellites", *Science* **307**, 1226–1236.
[42] *http://photojournal.jpl.nasa.gov/catalog/PIA07716*
[43] See reference [2]; also Farmer, A., Goldreich, P., 2006, "Understanding the behavior of Prometheus and Pandora", *Icarus* **180**, 403–411.
[44] Cuzzi, J. N., Burns, J. A., 1988, "Charged particle depletion surrounding Saturn's F ring: Evidence for a moonlet belt", *Icarus* **74**, 284–324; Barbara, J., Esposito, L. W., 2002, "Moonlet collisions and the effects of tidally modified accretion in Saturn's F ring", *Icarus* **160**, 161–171; McGhee *et al.* [5].
[45] Murray, C. D., Chavez, C., Buerle, K., Cooper, N., Evans, M. W., Burns, J. A., Porco, C. C., 2005, "How Prometheus creates structure in Saturn's F ring", *Nature* **437**, 1326–1329.
[46] Smith, B. A., Soderblom, L. A., Batson, R. M., Bridges, P. M., Inge, J. L., Masursky, H., Shoemaker, E. M., Beebe, R. M., Boyce, J. M., Briggs, G. A., Bunker, A. S., Collins, S. A., Hansen, C. J., Johnson, T. V., Mitchell, J. L., Terrile, R. J., Cook, A. F. II, Cuzzi, J. N., Pollack, J. B., Danielson, G. E., Ingersoll, A. P., Davies, M. E., Hunt, G. E., Morrison, D., Owen, T. C., Sagan, C., Veverka, J., Strom, R. G., Suomi, V. E., 1982, "A new look at the Saturn system: The Voyager 2 images", *Science* **215**, 504–537.
[47] Cuzzi, J. N., Scargle, J. D., 1985, "Wavy edges suggest moonlet in Encke's gap", *Astrophysical Journal* **292**, 276–290; Showalter, M. R., Cuzzi, J. N., Marouf, E. A., Esposito, L. W., 1986, "Moonlet wakes and the orbit of the Encke gap moonlet", *Icarus* **66**, 297–323.
[48] Showalter, M. R., 1991, "Visual detection of 1981S13, Saturn's eighteenth satellite, and its role in the Encke gap", *Nature* **351**, 709–713.
[49] Showalter *et al.* [47].
[50] *http://saturn.jpl.nasa.gov/multimedia/images/image-details.cfm?imageID = 1521*
[51] Lewis, M. C., Stewart, G. R., 2005, "Expectations for Cassini observations of ring material with nearby moons", *Icarus* **178**, 124–143.
[52] Lissauer, J. J., Shu, F. H., Cuzzi, J. N., 1981, "Moonlets in Saturn's rings?", *Nature* **292**, 707–711; Henon, M., 1981, "A simple model of Saturn's rings", *Nature* **293**, 33–35.
[53] Seiß, M., Spahn, F., Sremcêvić, M., Salo, H., 2005, "Structures induced by small moonlets in Saturn's rings: Implications for the Cassini Mission", *Geophysical Research Letters* **32**, 1–4 (Cite ID L11205).
[54] Tiscareno, M. S., Burns, J. A., Hedman, M. M., Porco, C. C., Weiss, J. W., Dones, L., Richardson, D. C., Murray, C. D., 2006, "100-metre-diameter moonlets in Saturn's A ring from observations of 'propeller' structures", *Nature* **440**, 648–650.
[55] See various chapters in *Planetary Rings*, 1984, edited by Greenberg, R. and Brahic, A.
[56] Poulet, F., Sicardy, B., 2001, "Dynamical evolution of the Prometheus–Pandora system", *Monthly Notices of the Royal Astronomical Society* **322**, 343–355.
[57] Humes, D. H., 1980, "Results of Pioneer 10 and 11 meteoroid experiments: Interplanetary and near-Saturn", *Journal of Geophysical Research* **85**, 5841–5852; Grün, E., Zook, H. A., Festig, H., Giese, R. H., 1985, "Collisional balance of the meteoritic complex", *Icarus* **62**, 244–272; see also reference [58].
[58] Cuzzi, J. N., Estrada, P. R., 1998., "Compositional evolution of Saturn's rings due to meteoroid bombardment", *Icarus* **132**, 1–35.
[59] Srama, R., Altobelli, N., Kempf, S., 2005 and 2006, personal communications.
[60] Kempf, S., 2006, personal communication.
[61] Showalter, M. R., 1998, "Detection of centimeter-sized meteoroid impact events in Saturn's F ring", *Science* **282**, 1099–1100.

[62] McGhee, C. A., French, R. G., Dones, L., Cuzzi, J. N., Salo, H. J., Danos, R., 2005, "HST observations of spokes in Saturn's B ring", *Icarus* **173**, 508–521.

[63] Nitter, T., Havnes, O., Melandsø, F., 1998, "Levitation and dynamics of charged dust in the photoelectron sheath above surfaces in space", *Journal of Geophysical Research* **103**, 6605–6620; Mitchell, C. J., Horányi, M., Havnes, O., Porco, C. C., 2006, "Saturn's spokes: Lost and found", *Science* **311**, 1587–1589; Farrell, W. M., Desch, M. D., Kaiser, M. L., Kurth, W. S., Gurnett, D. A., 2006, "Changing electrical nature of Saturn's rings: Implications for spoke formation", *Geophysical Research Letters* **33**, Issue 7, Cite ID L07203.

[64] Goertz, C. K., Morfill, G., 1983, "A model for the formation of spokes in Saturn's rings", *Icarus* **53**, 219–229; Morfill, G., Goertz, C. K., 1983, "On the evolution of Saturn's 'spokes': Theory", *Icarus* **53**, 230–235; Morfill, G. E., Festig, H., Grün, E., Goetz, C. K., 1983, "Some consequences of meteoroid impacts on Saturn's rings", *Icarus* **55**, 439–447.

[65] Farmer, A. J., Goldreich, P., 2005, "Spoke formation under moving plasma clouds", *Icarus* **179**, 535–538; Morfill, G., Thomas, H. M., 2005, "Spoke formation under moving plasma clouds: The Goertz–Morfill model revisited", *Icarus* **179**, 539–542.

[66] Tagger, M., Henriksen, R. N., Pellat, R., 1991, "On the nature of the spokes in Saturn's rings", *Icarus* **91**, 297–314.

[67] Yanagisawa, M., Kisaichi, N., 2002, "Lightcurves of 1999 Leonid impact flashes on the Moon", *Icarus* **159**, 31–38; Ortiz, J. L., Quesada, J. A., Aceituno, J., Aceituno, F. J., Bellot Rubio, L. R., 2002, "Observation and interpretation of Leonid impact flashes on the Moon in 2001", *Astrophysical Journal* **576**, 567–573.

[68] Chambers, L. S., Cuzzi, J. N., Asphaug, E., Colwell, J. E., Sugita, S., 2006, "Meteoroid impacts into Saturn's rings", *Icarus*, submitted.

[69] Gurnett, D., Kurth, W., Hospodarsky, G., Persoon, A., Desch, M., Farrell, W., Kaiser, M., Goetx, K., Cecconi, G., Lecacheux, A., Zarka, P., Harvey, C., Louarn, P., Canu, P., Cornilleau-Wehrlin, N., Galopeau, P., Roux, A., Fischer, G., Ladreiter, H., Rucker, H., Alleyne, H., Bostrom, R., Gustafsson, G., Wahlund, J., Pedersen, A., 2004, "An overview of Cassini radio, plasma wave, and Langmuir probe observations in the vicinity of Saturn", Fall AGU Meeting, abstract #P54A-02.

[70] Doyle *et al.* [9].

[71] Durisen, R. H., Bode, P. W., Cuzzi, J. N., Cederbloom, S. E., Murphy, B. W., 1992, "Ballistic transport in planetary ring systems due to particle erosion mechanisms. II: Theoretical models for Saturn's A- and B-ring inner edges", *Icarus* **100**, 364–393.

[72] Clark, R. N., Brown, R., Baines, K., Bellucci, G., Bibring, J., Buratti, B., Capaccioni, F., Cerroni, P., Combes, M., Coradini, A., Cruikshank, D., Drossart, P., Filacchione, G., Formisano, V., Jaumann, R., Langevin, Y., Matson, D., McCord, T., Mennella, V., Nelson, R., Nicholson, P., Sicardy, B., Sotin, C., Curchin, J., Hoefen, T., 2005, "Compositional mapping of surfaces in the Saturn system with Cassini VIMS: The role of water, cyanide compounds and carbon dioxide", AGU Fall Meeting 2005, abstract #P22A-02.

[73] Northrop, T. D., Connerney, J. E. P., 1987, "A micrometeorite erosion model and the age of Saturn's rings", *Icarus* **70**, 124–137.

[74] Lissauer, J. J., Squyres, S., Hartmann, W., 1988, "Bombardment history of the Saturn system", *Journal of Geophysical Research* **93**, 13776–13804.

[75] Dones, L., 1991, "A recent cometary origin for Saturn's rings?", *Icarus* **92**, 194–203.

[76] Brown, R. H., Baines, K. H., Bellucci, G., Buratti, B. J., Capaccioni, F., Cerroni, P., Clark, R. N., Coradini, A., Cruikshank, D. P., Drossart, P., Formisano, V., Jaumann, R., Langevin, Y., Matson, D. L., McCord, T. B., Mennella, V., Nelson, R. M., Nicholson,

P. D., Sicardy, B., Sotin, C., Baugh, N., Griffith, C. A., Hansen, G. B., Hibbitts, C. A., Momary, T. W., Showalter, M. R., 2006, "Observations in the Saturn system during approach and orbital insertion, with Cassini's visual and infrared mapping spectrometer (VIMS)", *Astronomy and Astrophysics* **446**, 707–716.

[77] Brown, M. E., Schaller, E. L., Roe, H. G., Rabinowitz, D. L., Trujillo, C. A., 2006, "Direct Measurement of the size of 2003UB313 from the Hubble Space Telescope", *Astrophysical Journal* **643**, L61–L63.

## 10.10 PICTURES AND DIAGRAMS

Figure 10.1 *http://saturn.jpl.nasa.gov/multimedia/images/image-details.cfm?imageID=1943*
Figure 10.2 *http://samadhi.jpl.nasa.gov/art/flybys.html*
Figure 10.3 Cassini Project, JPL, and NASA.
Figure 10.4 *http://photojournal.jpl.nasa.gov/catalog/PIA06142* for nice color version; stretch by J. Cuzzi.
Figure 10.5 *http://photojournal.jpl.nasa.gov/catalog/PIA06529*
Figure 10.6 Image obtained from the NASA PDS website and processed by J. Cuzzi.
Figure 10.7 Porco, C. C. *et al.*, 2005, "Cassini imaging science: Initial results on Saturn's rings and small satellites", *Science* **307**, 1226–1236; *http://photojournal.jpl.nasa.gov/catalog/PIA06096*, and *http://photojournal.jpl.nasa.gov/catalog/PIA06093*
Figure 10.8 *http://photojournal.jpl.nasa.gov/catalog/PIA06543* and *http://photojournal.jpl.nasa.gov/catalog/PIA06535*
Figure 10.9 Porco, C. C. *et al.* (2005), *loc. cit.*
Figure 10.10 *http://photojournal.jpl.nasa.gov/catalog/PIA06540*
Figure 10.11 Adapted from Salo, H. *et al.*, 2004, "Photometric modeling of Saturn's rings. II. Azimuthal asymmetry in reflected and transmitted light", *Icarus* **170**, 70–90.
Figure 10.12 *http://photojournal.jpl.nasa.gov/catalog/PIA06606*
Figure 10.13 *http://photojournal.jpl.nasa.gov/catalog/PIA07556*
Figure 10.14 Esposito, L. W. *et al.*, 2005, "Ultraviolet imaging spectroscopy shows an active Saturnian system", *Science* **307**, 1251–1255.
Figure 10.15 *http://photojournal.jpl.nasa.gov/catalog/PIA07718*
Figure 10.16 *http://photojournal.jpl.nasa.gov/catalog/PIA07714*
Figure 10.17 From Nicholson, P. *et al.*, 2006, "SOI observations of Saturns rings", *Icarus*, submitted.
Figure 10.18 Courtesy VIMS team and Cassini project; soon to be released.
Figure 10.19 *http://photojournal.jpl.nasa.gov/catalog/PIA06653*
Figure 10.20 *http://photojournal.jpl.nasa.gov/catalog/PIA06175*
Figure 10.21 *http://photojournal.jpl.nasa.gov/catalog/PIA07872*
Figure 10.22 *http://photojournal.jpl.nasa.gov/catalog/PIA07960*
Figure 10.23 Wallis, B. *et al.*, 2005, "CIRS observations of a thermal enhancement near zero phase in Saturn's rings"; AGU Fall Meeting 2005, abstract #P33B-0250; Spilker, L. *et al.*, 2006, "Cassini CIRS observations of thermal differences in Saturn's main rings with increasing phase angle", *37th Lunar and Planetary Science Conference*, abstract #2299.
Figure 10.24 Spilker, L. *et al.*, 2005, "Cassini CIRS observations of Saturn's rings", *36th Lunar and Planetary Science Conference*, abstract #1912.

Figure 10.25 Waite *et al.*, 2005, "Oxygen ions observed near Saturn's A ring", *Science* **307**, 1260–1262.
Figure 10.26 *http://photojournal.jpl.nasa.gov/catalog/PIA06143* and *http://photojournal.jpl.nasa.gov/catalog/PIA07523*
Figure 10.27 Porco, C. C. *et al.*, 2005, *loc. cit.*; *http://photojournal.jpl.nasa.gov/catalog/PIA07716*
Figure 10.28 *http://photojournal.jpl.nasa.gov/catalog/PIA07632*
Figure 10.29 *http://photojournal.jpl.nasa.gov/catalog/PIA06099*; note, however, that the image as displayed in the various Cassini websites is incorrectly flipped by 180 degrees vertically.
Figure 10.30 *http://photojournal.jpl.nasa.gov/catalog/PIA06237*
Figure 10.31 Tiscareno, M. *et al.*, 2006, "100-metre-diameter moonlets in Saturn's A ring from observations of 'propellor' structures", *Nature* **440**, 648–650
Figure 10.32 Cuzzi, J. N., Estrada, P. R., 1998, "Compositional evolution of Saturn's rings due to meteoroid bombardment", *Icarus* **132**, 1–35.
Figure 10.33 *http://photojournal.jpl.nasa.gov/catalog/PIA07731*

# 11

# Comparative planetology of the giant planet ring systems

## 11.1 INTRODUCTION

The Voyager 1 and 2 encounters of Jupiter, Saturn, Uranus, and Neptune provided us for the first time detailed views of the ring systems of all four planets, and, for the first time, in color. The images and other data from Voyager were a snapshot in time of rings whose features vary over intervals of years, and in some cases over months or even days. Occasional observations of these ring systems by the Hubble Space Telescope and other Earth-based observatories provided the first evidence of this variability, and data from the Cassini Orbiter have provided ample evidence of that variability in Saturn's ring system.

In prior chapters, we have made an effort to provide with every major ring characteristic an estimate of similar conditions within the ring systems of the other gas giant planets. Because much can be learned about each ring system from such "comparative planetology of the giant planet ring systems", we have chosen to gather together in one chapter all such comparisons. These comparisons are organized below under the topics of ring dimensions, ring azimuthal structure, ring-particle characteristics, gravitational interactions of rings with satellites, electromagnetic interactions of rings with magnetic fields, ring creation, evolution and age, and unanswered questions. Most of the comparisons will be in text form, although tables and figures will be used where they tell the story more clearly.

## 11.2 RING DIMENSIONS

With the exception of the only known blue-tinted rings (Saturn's E ring and Uranus's recently discovered 2003U1R ring, both of which appear to be composed of water-ice particles derived from satellites Enceladus and Mab, respectively), all of the known planetary rings are found within 3.2 planetary radii of the centers of their respective planets (Figure 11.1). Well exterior to these rings whose radial positions and dimensions are catalogued in Table 11.1, Uranus and Neptune have dust disks detected by plasma pulses created as they impacted Voyager 2 during and near ring-plane crossing times. All of the rings and dust disks are centered near the equators of their respective planets, or, more precisely, near their *local Laplace planes* [1], which differ from the equatorial planes due to the combined gravitational effects of the Sun and of large satellites in inclined orbits. Exceptions to this general rule include several of the narrow rings of Uranus which have measurable non-zero inclinations. The inner and outer Gossamer rings of Jupiter derive their material from satellites with inclined orbits, but the ring material is rapidly spread into two disks with zero inclination, a thickness equivalent to the maximum north–south excursions of the source satellites (Amalthea and Thebe), and a higher spatial density of ring particles near the upper and lower boundaries of the two disks. Only a handful of the narrow rings of Uranus and some of the ringlets within gaps in Saturn's rings have non-zero eccentricities, and except for the Uranus Epsilon ring (with an eccentricity of about 0.008), none of these eccentricities exceed 0.002. The outer edges of Saturn's A and B rings are not eccentric, but they are also non-circular. The A ring has seven lobes, and the B ring has two lobes.

## 11.3 RING AZIMUTHAL STRUCTURE

If one discounts the eccentric shape and corresponding width variations of Uranus's narrow rings, only the Saturn and Neptune ring systems show marked azimuthal variations in ring properties. For the Neptune ring system, the primary azimuthal structure is associated with the ring arcs in the Adams ring. As discussed in Chapter 8, the Adams ring arc structure has changed substantially since the Voyager 2 encounter with Neptune in 1989. Saturn's F ring and, to a lesser extent, the G ring show azimuthal variations indicative of the possible presence of ring parent bodies. Several of the narrow rings and ringlets seem to exhibit clumpiness. The enigmatic radial spokes in the outer half of Saturn's B rings are another example of azimuthal ring structure. They will be discussed again in Section 11.6.

## 11.4 RING-PARTICLE CHARACTERISTICS

The Jupiter ring particles seem to be composed of silicate dust. There may be a few larger particles in the two Gossamer rings, but the Main ring and Halo ring particles are predominantly micrometer and sub-micrometer in radius. Saturn's rings appear to be mainly water ice, with the possible exception of the G ring. The percentage of non-water constituents is somewhat larger in the C ring and Cassini Division than in the A and B rings. Micrometer-sized particles appear to dominate the D and E rings and the B-ring spokes and are present in smaller amounts in the other Saturn rings. The reflectivity of particles in the A and B rings is close to 60%; that of the C ring and Cassini Division is between 30% and 40%. In marked contrast to the reflective

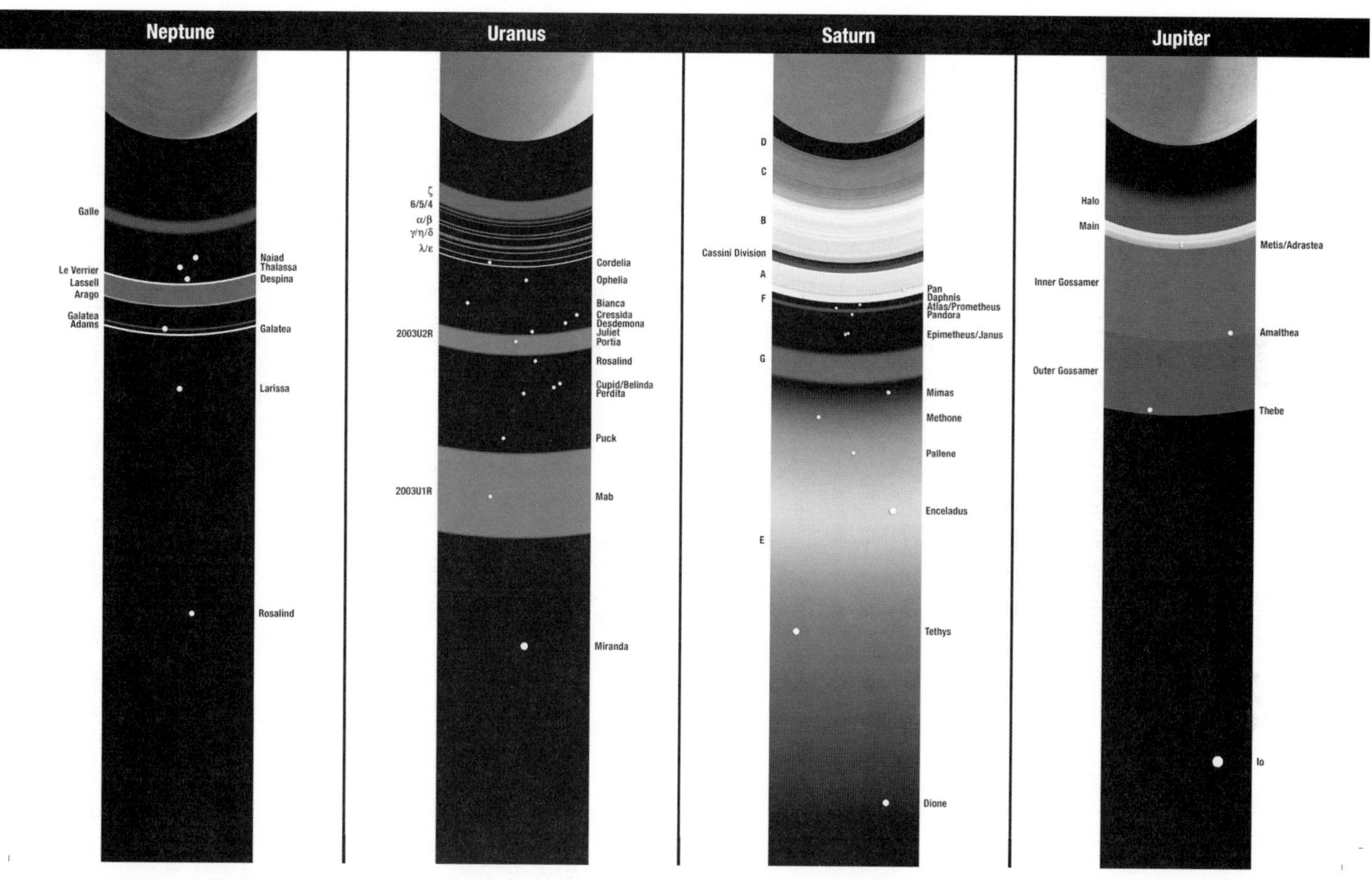


**Figure 11.1.** Diagram, scaled to the radius of each planet, of the four giant planet ring systems. Satellite names are given to the right of each ring system; ring names are labeled to the left of each ring system. For each planet, the Roche limit, the radius within which satellites are broken up by unbalanced tidal forces from the planet, lies between about 2 and 2.5 planetary radii.

**Table 11.1.** Radial dimensions of solar system rings.

| Ring region | Distances (km) | Distances ($R_P$) | Widths (km) |
|---|---|---|---|
| *Jupiter* | | | |
| Halo | 100,000–122,000 | 1.40–1.71 | 22,000 |
| Main | 122,000–129,200 | 1.71–1.81 | 7,200 |
| Inner Gossamer | 129,200–182,000 | 1.81–2.55 | 52,800 |
| Outer Gossamer | 182,000–224,900 | 2.55–3.15 | 42,900 |
| *Saturn* | | | |
| D | 66,900–74,510 | 1.11–1.235 | 7,610 |
| C | 74,510–92,000 | 1.235–1.525 | 17,490 |
| B | 92,000–17,580 | 1.525–1.949 | 25,580 |
| Cassini Division | 117,580–122,170 | 1.949–2.025 | 4,590 |
| A | 122,170–136,780 | 2.025–2.267 | 14,610 |
| F | 140,180 | 2.324 | 30 to 500 |
| G | 162,000–175,000 | 2.69–2.90 | 13,000 |
| E | 181,000–483.000 | 3.0–8.0 | 302,000 |
| *Uranus* | | | |
| $\zeta$ (Zeta) | 37,850–41,350 | 1.48–1.62 | 3,500 |
| 6 (Six) | 41,873.2 | 1.636 89 | 1 to 3 |
| 5 (Five) | 42,234.8 | 1.652 44 | 2 to 3 |
| 4 (Four) | 42,570.9 | 1.665 59 | 2 to 3 |
| $\alpha$ (Alpha) | 44,718.4 | 1.749 62 | 4 to 10 |
| $\beta$ (Beta) | 45,661.0 | 1.786 50 | 5 to 11 |
| $\eta$ (Eta) | 47,175.9 | 1.845 77 | 1 to 2 |
| $\gamma$ (Gamma) | 47,626.9 | 1.863 41 | 1 to 4 |
| $\delta$ (Delta) | 48,300.1 | 1.889 75 | 3 to 7 |
| $\lambda$ (Lambda) | 50,023.9 | 1.957 19 | 2 to 3 |
| $\varepsilon$ (Epsilon) | 51,149.3 | 2.001 23 | 20 to 96 |
| 2003U2R | 65,400–69,200 | 2.56–2.71 | 3,800 |
| 2003U1R | 89,200–106,200 | 3.49–4.16 | 17,000 |
| *Neptune* | | | |
| Galle | 40,900–42,900 | 1.66–1.74 | 2,000 |
| Le Verrier | 53,200 | 2.15 | ~110 |
| Lassell | 53,200–57,200 | 2.15–2.31 | 4,000 |
| Arago | 57,200 | 2.31 | <100 |
| (Galatea) | 61,950 | 2.50 | <100 |
| Adams | 62,932 | 2.54 | 15 to 50 |

$R_P$ = Radial range in relevant planetary radii. Note that distances are from center of each planet.

particles in Saturn's rings, the rings of Uranus are extremely dark. Particles in the Epsilon ring (and possibly all of the narrow Uranian rings) are some of the darkest bodies known, reflecting 3.5% or less of the sunlight incident upon them. Among naturally occurring materials, only elemental carbon is that dark; it is therefore presumed that at least the surfaces of ring particles in the narrow rings of Uranus are covered with carbon. The two recently discovered rings of Uranus do not follow that general mold. 2003U1R had a distinct blue tint, making it similar only to Saturn's E ring in that respect; its composition may also be water-ice particles blasted from the surface of Mab, which is centrally located in that ring. 2003U2R is redder and more reminiscent of Saturn's G ring, presumed to be mainly silicate in composition. Neptune's narrow rings and arcs are also very dark, again a possible indication of carbon composition, at least of the ring-particle surfaces. The narrow rings of Uranus, the Lambda ring excepted, seem remarkably void of particles smaller than a few centimeters in radius, although a dust band consisting of micrometer-sized particles which permeates all the space between the narrow rings was clearly seen in a single high-phase-angle image obtained by Voyager 2. Radial structure within that dust band is plentiful; the causes for that structure are unknown, as are the mechanisms for confining the narrow rings of Uranus. One possibility is that both the confinement mechanism for particles within the narrow rings and some or all of the radial structure in the dust band are due to small satellites that have not yet been discovered.

## 11.5 GRAVITATIONAL INTERACTIONS OF RINGS WITH SATELLITES

Orbital resonances between rings and satellites are responsible for features in all four ring systems, with the possible exception of the Jupiter ring system. These resonances occur when the orbital period of a given satellite and the orbital period of the ring particles it affects have a simple, generally low-number fractional relationship to one another.

The outer edge of Saturn's A ring occurs at a point where the ring particles complete seven orbits of Saturn for every six orbits of the satellite Janus; that outer ring edge is consequently seven-lobed in shape. Similarly, ring particles at the outer edge of Saturn's B ring complete two orbits of Saturn for each orbit of Mimas, and that ring edge is two-lobed. These strong resonances are known as 7:6 and 2:1 inner Lindblad resonances, respectively. Ring particles at the inner and outer edges of Uranus's Epsilon ring are prevented from escaping that ring by a 24:25 outer Lindblad resonance of Cordelia and a 14:13 inner Lindblad resonance of Ophelia, respectively. A 42:43 outer Lindblad resonance of Galatea may not only radially confine particles in Neptune's Adams ring, but the 42 lobes caused by that orbital resonance may additionally impede longitudinal spreading of material in the ring arcs from one 8.57° cell to the next.

Three other types of gravitational interactions between ring particles and satellites are noted. Density waves are initiated by orbital resonances that are somewhat weaker than the ring-confinement mechanisms of the previous paragraph. More than 50 such density waves have been identified in Saturn's A and B rings. A tentative identification was also made of a density wave in Uranus's Delta ring. Vertical tugs by Mimas, whose orbit is inclined to Saturn's equator by 1.51°, cause so-called bending waves or local corrugations in rings. Thus far, the only identified bending waves in any of the planetary ring systems are located in Saturn's A ring and are caused by Mimas. The third type of interaction occurs when a satellite is imbedded within an otherwise continuous ring. If the satellite is massive enough, it can actually clear a gap, such as the Encke gap in Saturn's A ring, which is caused by the satellite Pan. In addition, Pan also gives rise to satellite wakes at the inner and outer edges of the Encke gap. Because of differential rotation, the wakes on the inner edge of the Encke gap precede Pan in its orbit; those at the outer edge trail Pan. As discussed in Chapter 10, a new moonlet (Daphnis) found in the Keeler gap in the outer A ring also creates wakes along the edge of that gap. Furthermore, there are a number of other structures at the limits of visibility in the A ring that may be wakes of even tinier moons. It now seems likely that all of the empty gaps seen in Saturn's rings are due to clearing by moonlets not yet discovered.

## 11.6 ELECTROMAGNETIC INTERACTIONS OF RINGS WITH MAGNETIC FIELDS

Each of the giant planets generates a magnetic field that is populated with ions and electrons which are swept along with that field as it rotates at the same rate as the interior of the planet. The planetary ring systems occupy that portion of space where that magnetic field is generally stronger, but the ring particles orbit at speeds dictated by Kepler's laws, not at the planet's rotation rate. As the ions and electrons "frozen" in the magnetic field sweep past the ring particles, there are two results. First, much of the charged particle population in that part of the magnetosphere is absorbed by the ring particles, thereby reducing the number density of charged particles. That reduction has at times been noted before the ring itself was visually detected, as in the case of Pioneer 11's discovery of Saturn's G ring. Second, if the ring particles are small enough, their charge-to-mass ratio can become large enough that electromagnetic forces begin to be stronger than gravitational forces, and strange results can occur. Such is the case for Jupiter's Halo ring, where the sub-micrometer ring particles are lifted vertically out of the ring plane, especially at radii which correspond to low-order Lorentz resonances. Such resonances occur where the ratio between the ring-particle orbital period and the rotation period of the planetary magnetic field are related by low-order integers, such as 4:3, 3:2, 2:1, 1:2, 2:3, etc. The outer edge of the Halo ring is at the 3:2 Lorentz resonance radius for Jupiter; the inner edge is at the 2:1 Lorentz resonance radius. As particles in Jupiter's Main ring migrate slowly planetward, they accumulate charge, are broken into smaller and smaller pieces, and then encounter the 3:2 Lorentz resonance, where the magnetic field begins to carry them vertically out of Jupiter's equatorial plane. As the particles continue to migrate inward, they eventually reach the stronger 2:1 Lorentz resonance and are removed from the ring system entirely.

For this mechanism to work effectively, the magnetic field needs to be tilted with respect to the rotation axis. In the case of Saturn, the tilt of the magnetic field is very nearly 0 (Jupiter's is about 10°), and little evidence for vertical accelerations is seen. However, in the outer B ring, the orbital period is very nearly the same as the rotation period of the planet. As clouds of dusty debris are created by meteoroid impacts with ring particles, those micrometer-sized particles can become electrically charged, either by the freeing of electrons through the energy imparted from sunlight (photoelectric effect) or by collisions with and absorption of electrons and ions in the adjacent magnetosphere. The radial "spokes" formed by such collisions often follow the rotation rate of the magnetic field for a time, maintaining their radial structure until they lose their charge and are destroyed by Keplerian shear. Keplerian shear occurs because ring particles closer to the planet orbit the planet at faster speeds and in less time that those slightly further out.

Uranus's magnetic field is highly tilted to its rotation axis. It is possible that the consequent frequent sweeping of the magnetic field through the rings may, over time, remove a major fraction of the dust and small particles from the ring system. In such a scenario, the dust band seen in the single high-phase-angle image may be dusty debris from relatively recent impacts, and that the dust had not yet acquired sufficient charge to be strongly affected by the magnetic field.

The highly tilted magnetic field of Neptune ought to have a similar effect on that planet's ring system, but sunlight levels are lower and the magnetosphere there is populated by a much lower spatial density of charged particles. Whatever the reason, the rings do not seem as devoid of small particles as those of Uranus.

Another electromagnetic effect operative in all four systems is Poynting–Robertson drag, or Poynting–Robertson effect. Ring particles, especially tiny particles, absorb sunlight from the direction of the Sun and re-emit it in all directions, creating a net acceleration on the particles. This acceleration eventually causes tiny particles to spiral into the planet's atmosphere. The presence of Poynting–Robertson drag is one of the reasons that the present ring systems are believed to be younger than the age of the solar system, perhaps even less than 1% of that age.

## 11.7 RING CREATION, EVOLUTION, AND AGE

The Jupiter ring system is perhaps more completely understood at present than any of the other three ring systems. With its enormous mass, Jupiter's gravitational pull attracts a large number of meteoroids and micrometeoroids. The four inner moons (Metis, Adrastea, Amalthea, and Thebe) appear to be almost constantly bombarded by these objects, and the debris from this bombardment is almost certainly the source of the ring material for the known rings. Much of the material from the larger moonlets Amalthea and Thebe falls back to their surfaces, creating a soil covering (regolith) and slowly burying evidence of older impacts. However, some of it remains in orbit around Jupiter. The more vertically extended Outer (or Thebe) Gossamer ring comes from Thebe, whose inclination is slightly higher than that of Amalthea, which feeds the Inner (or Amalthea) Gossamer ring. As the debris is broken into smaller and smaller pieces by continuing bombardment, they begin to reach sizes small enough that Poynting–Robertson drag begins to move them inward toward Jupiter, thereby creating extended disks, more heavily populated near their upper and lower surfaces because the orbital motion of Thebe and Amalthea causes these moonlets to spend more time near those surfaces. In the meantime, the smaller moonlets Metis and

Adrastea also undergo heavy bombardment, but their smaller masses do not re-attract as much of the resultant debris, and a denser (and vertically thinner) Main ring is formed. Like the Gossamer rings, its particles are broken into smaller and smaller pieces by continued bombardment, but because the Main ring is more densely populated, the resultant particles begin to attain charge from Jupiter's magnetospheric plasma. For most of the particles, Poynting–Robertson drag remains the dominant perturbing force for these small particles, but when they reach the radial distance of the 3:2 Lorentz resonance, electromagnetic forces from Jupiter's inclined magnetic field begin to act repeatedly on the particles, vertically forcing them to orbits that are above and below the equator, and a Halo ring is formed. Poynting–Robertson drag continues to push the particles planetward, and they continue to accumulate electrical charge. When they eventually reach the radial distance of the 2:1 Lorentz resonance, electromagnetic forces completely dominate over gravitational forces, and the tiny particles are either ejected from the Jupiter system or they are propelled into the atmosphere, where they become part of the planetary mass.

The relative simplicity and low mass of the Uranus and Neptune rings make their explanations somewhat easier to fathom, although we have not yet identified either their sources nor their confinement mechanisms (other than a likely confinement mechanism for the Uranus Epsilon ring). The source bodies may be buried within the rings themselves, hidden by the ring particles, perhaps in much the same way as for Saturn's F ring after the Voyager encounters. In the case of the recently discovered blue-tinged ring of Uranus, 2003U1R, the source of the ring is very likely Mab, which is likely an icy body, bombardment of which has created a relatively unconstrained ring of tiny ice particles. The unnamed ring that extends around (or partially around) the orbit of Neptune's Galatea may be of the same nature as the narrow rings of Uranus, but younger, less well developed, and emanating from a somewhat larger satellite. They are also reminiscent of the tiny moons found in the Encke and Keeler gaps in Saturn's A ring, forming complete or partial ringlets within those gaps. If, as suggested by the occultation data, the Uranus rings are deficient in small particles, it is possible that the wildly tilted and offset magnetic field of Uranus has aided in sweeping them almost clean of small particles. The lower levels of sunlight and sparser magnetospheric plasma population at Neptune might combine to reduce significantly the amount of charging of small ring particles and thereby explain why the Neptune rings still retain a more substantial population of tiny ring particles than Uranus.

The story for Saturn's rings is even more enigmatic. The extended E-ring is likely generated by water-ice eruptions from the active satellite Enceladus. The particles move inward and outward from Enceladus, and the tiny particles are not nearly as tightly constrained to the equatorial plane of Saturn as are the inner rings. As the ring thickness increases with distance from the planet, its spatial density of particles decreases, and although the ring may extend nearly to the orbit of Titan, it has not been visually detected beyond about half that radial distance.

The F and G rings are probably generated from bombardment of and mutual collisions between tiny moons resident within these rings. The Cassini Orbiter has already discovered three such moonlets orbiting within the F ring. Prometheus and Pandora, which at one time were thought to shepherd and prevent radial spreading of

F-ring particles, probably assist in gravitationally stirring up the ring particles and perhaps inducing additional collisions, but they apparently do very little to constrain radial spreading of the F ring.

The source of material in Saturn's D ring may be dust particles being moved planetward from the A, B, and C rings primarily by Poynting–Robertson drag, but like Uranus's dusty disk of particles, seen only at high phase angles, there is as yet no workable explanation for the very organized radial structure which exists in these two tenuous rings.

Saturn's A, B, and C rings contain about the same amount of mass as the 400-km diameter moon Mimas. Dynamic processes occurring within these rings argue for a lifetime that is much shorter than the 4.5-billion-year age of the solar system, perhaps less than 1% of that age. They are therefore unlikely to be unorganized material left over from the time of Saturn's formation. Furthermore, their intrinsic brightness bespeaks a relatively recent origin. The most likely explanation, particularly for the more massive A and B rings, is the catastrophic breakup of a parent body about the size of Mimas. However, the spectrum of the rings is unlike that of any of Saturn's satellites; the rings appear significantly redder. The origin of the parent body (or bodies) is therefore unknown, with nearly equal numbers of proponents for an exterior origin (perhaps from the Kuiper Belt beyond Neptune) or an origin from a body formed within the Saturn system. The source of the parent body is only one of the major problems facing ring scientists. Neither the unbalanced tidal forces experienced by a Mimas-sized or larger body (to allow for ring-particle capture inefficiencies), nor collisions with Sun-orbiting bodies within the past 100,000,000 to 500,000,000 years or so (when the flux of large impactors is thought to have already been depleted), are capable of easily breaking up the parent body into ring-particle-sized debris.

## 11.8 UNANSWERED QUESTIONS

It has been very nearly 400 years since Saturn's rings were first sighted. The other three giant planet ring systems have all been discovered since 1977, a scant 30 years ago. While these more recently discovered ring systems are far from completely understood, they present less of a challenge to ring scientists than the A and B rings of Saturn. Mechanisms for generating and shaping and replenishing those ring systems can be envisioned, even in the absence of specific observations of the satellites needed to confine the narrow rings of Uranus and Neptune. Unless we have badly misjudged the structures and dynamics of these rings, future observations from more capable Earth-based observatories or from properly equipped man-made probes sent to these giant planets will eventually find large numbers of moonlets, some in precisely the right orbits to gravitationally constrain or otherwise shape the known or yet-to-be-discovered rings. While we are beginning to understand these latecomer rings, we have not yet accomplished much more than to scratch the surface of understanding Saturn's ring system.

Many of the "big picture" problems in Saturn's rings that faced us before the 1980 and 1981 Voyager encounters remain unanswered. For example, what is the origin of the rings, and why are they still unique among known planetary ring systems? Cassini has provided enough insight to enable us to start asking such questions on a more sophisticated level and with greater clarity. However, the new details also require us to understand the present structure better before we can delve into the question of ring origin.

At the most basic level, as we have described in the foregoing chapters, the story of the rings is the story of gravitational interactions between ring material, nearby satellites, and the parent planet. In all four gas giant systems, one finds more numerous, more closely spaced, and smaller satellites as one approaches their parent planet, merging into ring systems very close to the planet. Very small satellites (moonlets) are even found sprinkled within the ring systems themselves. This basic characteristic is related to the so-called Roche limit of the planet, discussed in Section 1.5.

Most of the interaction between Saturn's rings and nearby satellites is tied to the physics of resonant spiral waves, discussed in Section 10.3 and elsewhere in this book. These spiral wave patterns are each driven by repetitive forces from a particular satellite, forces which become significant near specific locations (known as orbital resonances) in the rings where the orbit period of the ring material is a simple fraction of the orbit period of the satellite. The tiny self-gravity of the ring material amplifies these perturbations by the satellite, and a wave is set up which has the shape of a tightly wrapped spiral, whose structure is intimately related to the physics which creates the arms of spiral galaxies.

Smaller, but closer, moonlets affect ring material in a related, but somewhat different manner. The moonlet may even be embedded within the rings. The influence of a nearby moonlet comes almost entirely as it passes by a section of a ring, and is often described as a gravitational impulse. The effect of these gravitational impulses on the ring is to create patterns which transfer angular momentum between the moonlet and the ring material, and in this case the ring material recedes from the moonlet, causing a nearly empty gap to form. As of this writing, two such moonlets have been identified; Pan is responsible for the Encke gap in the A ring, and a smaller moonlet by the name of Daphnis has cleared ring material from the narrower Keeler gap near the outer edge of the A ring.

Still smaller moonlets, all less than a kilometer in diameter, are apparently imbedded within the A and B rings. While these moonlets do not have enough mass to clear gaps within the rings and are not large enough to be resolved in images, they do create local disturbances which can be seen in the highest resolution images of the rings. In the same way that gap-clearing moonlets create nearby wakes at the edges of the gaps they clear, these smaller moonlets create short-lived wakes that move ahead of the moonlet on the planetward side of the moonlet and others that trail the moonlet on the outward side.

The ring particles, all orbiting Saturn at speeds faster than a rifle bullet, move only very slowly with respect to their neighbors; they are essentially flying in formation. The innumerable gentle collisions between these often densely packed ring particles make them behave in many ways like molecules in a fluid, with properties that can be described as compressibility, viscosity, pressure, and so forth. Just as the viscosity of a fluid determines how it flows, the viscosity of the rings determines how they evolve—it enters into all the physics of spiral waves, shepherding, ring spreading, etc. The ring viscosity is most closely related to the random velocity between particles. Determination of this important local property is of great importance, but it can only be determined indirectly, because it is highly unlikely that we will ever measure the millimeter-per-second random velocity of any individual ring particle directly. Some indirect determinations of the particle random velocities have been possible, as discussed in Section 10.3.

There are numerous other features in the ring systems whose structure, evolution, and/or history are not understood. The causes of the so-called irregular structure, found primarily in the inner half of Saturn's B ring, are not understood, although at least one somewhat esoteric explanation has been attempted [2]. The fine-scale radial structure in Saturn's D ring and the dusty ring disk of Uranus has not yet been satisfactorily explained. Most of the radial structure in the C ring is unexplained. For that matter, the unexplained features in the A and B rings far outnumber those for which a clear explanation exists. Certainly, some of the structure is due to small moonlets not yet discovered in the rings. At present, there is no definitive dividing line between embedded moonlets and large ring particles, and indeed there may be a continuum of sizes from Pan and Daphnis all the way down to sub-micrometer ring particles. It would be helpful to have a better understanding of ring-particle size distribution and interaction, especially in the optically thick A and B rings. It is clear that the remaining years of the Cassini Orbiter mission (including a likely extended mission) will provide enormous quantities of ring data that bear on the unknowns in Saturn's ring system; some of those data will shed light on the unanswered questions about structure, evolutionary processes, and ages of the ring system, and we will be much wiser about it than we are at this time. It is equally clear that, even after decades of analysis of Cassini ring data, there will remain mysteries in what is observed that will beg for more data, both observational and theoretical. Perhaps you, dear reader, will be one of those whose flash of insight will lead to a breakthrough in understanding these beautiful, but unfathomably complex jewels of the solar system.

## 11.9 NOTES AND REFERENCES

[1] The local Laplace plane, as applied to ring material, is the near-equatorial plane as altered by the gravitational effects of the Sun and large satellites (such as Saturn's Titan). For an inclined ring (such as the 4, 5, and 6 rings of Uranus), ring precession keeps the "average" plane of each ring at the local Laplace plane. Very near the giant planets, the local Laplace plane is essentially identical to the planet's equatorial plane. Farther out, the local Laplace plane generally departs from the equatorial plane in the direction of the planet's orbital plane.

[2] Tremain, S., 2003, "On the origin of irregular structure in Saturn's Rings", *Astronomical Journal* **125**, 894–901. In this article, the author suggests that ring-particle assemblies of up

to about 100 km in radial extent stick together and inhibit Keplerian shear from occurring; he further explains that the scale of the irregular structure is set by the competition between tidal forces from Saturn and the yield stress of these large, flat ice floes.

## 11.10 BIBLIOGRAPHY

Burns, J. A., 1999, "Planetary rings", Chapter 16 in *The New Solar System* (4th Edition), edited by Beatty, J. K., Petersen, C. C., Chaikin, A., Sky Publishing Corporation and Cambridge University Press, pp. 221–240.

Burns, J. A., Hamilton, D. P., Showalter, M. R., 2003, "Bejeweled worlds", *Scientific American* **13**, 74–83 (special edition entitled, "New Light on the Solar System").

Porco, C. C., Baker, E., Barbara, J., Beurle, K., Brahic, A., Burns, J. A., Charnoz, S., Cooper, N., Dawson, D. D., Del Genio, A. D., Denk, T., Dones, L., Dyudina, U., Evans, M. W., Giese, B., Grazier, K., Helfenstein, P., Ingersoll, A. P., Jacobson, R. A., Johnson, T. V., McEwen, A., Murray, C. D., Neukum, G., Owen, W. M., Perry, J., Roatsch, T., Spitale, J., Squyres, S., Thomas, P., Tiscareno, M., Turtle, E., Vasavada, A. R., Veverka, J., Wagner, R., West, R., 2005, "Cassini imaging science: Initial results on Saturn's rings and small satellites", *Science* **307**, 1226–1236.

## 11.11 PICTURES AND DIAGRAMS

Figure 11.1 Original artwork.

# 12

# Anticipated future observations of planetary ring systems

## 12.1 INTRODUCTION

For those of you who have stayed with us through this entire book, we sincerely hope that the journey has been enjoyable for you. It has been, in a sense, a labor of love for us to produce the book. All three of the authors have been intimately involved in the NASA space program for essentially the duration of our professional careers. Although only one of us is a recognized expert on planetary ring systems, as those of you who have browsed through the lists of authors in the cited references at the end of each chapter have undoubtedly already realized, all three of us have a love of the space program and a desire to communicate these findings to both general and scientific audiences.

In this final chapter, we glance toward the horizon to see what lies ahead in the way of observations and theoretical studies that may help scientists to further unravel the complexity of planetary ring systems. The realm of many-body dynamics is not limited to planetary ring studies, of course, and significant inputs can be expected from researchers in galactic studies and the other fields mentioned in Chapter 1.

One interesting implication for spiral structure in galaxies has come from a better understanding of density waves in Saturn's rings. Like water in waves at the seashore, individual ring particles (and, by analogy, stars in a galactic spiral arm) do not move at the same rate as the spiral density waves seen in the images. Density waves in a planetary ring move around the planet at the same rate as the satellite whose gravity generates the density wave; the individual ring particles circle the planet much faster. It is possible that stars in a spiral galaxy similarly circle the galactic center at velocities greater than that of the spiral arms in which they temporarily reside. If that were not so, differential rotation would eventually destroy both the spiral density waves in rings and the spiral arms in galaxies.

In a similar fashion, continued theoretical and observational ring studies (and studies of more extensive disks elsewhere in the solar system, the Milky Way Galaxy, and the universe) will with time bring more complete understanding of the dynamics of and other physical processes operating within celestial disks. That in turn will bring insights into disk origins and evolution, perhaps including the origin and evolution of the protoplanetary disk that once surrounded the Sun and from which its planets formed.

By far the most significant influx of new data within the next few years will be the continued data return of the Cassini Mission from Saturn orbit. As mentioned in Chapter 10, Saturn ring studies will continue to be an important part of the Cassini objectives and data return, especially during the next few years when Cassini moves from near-equatorial orbits to near-polar orbits, where better ring viewing will be available. Of particular interest will be the next Saturn equinox, where the rings will transit through zero tilt as seen from the Sun and the northern face of the rings will be illuminated by the Sun for the first time in nearly 15 years. That transition will occur on August 10, 2009, about a year into a Cassini extended mission (assuming an extended mission is approved).

## 12.2 FUTURE TELESCOPIC OBSERVATIONS AND THEORETICAL STUDIES

Normally, a Saturn ring-plane crossing is a time of intense ground-based telescopic observation. However, at the time of the next crossing of Saturn's ring plane across the Earth (September 4, 2009), Saturn will be at a poor position in Earth's sky, a scant 11° from the Sun. The following crossing of Saturn's ring plane through the Earth will be March 23, 2025, and the Earth and Saturn will again be on opposite sides of the Sun, although Saturn will be slightly farther from the Sun in Earth's sky than in 2009. Hence, no reasonably good Saturn ring-plane crossings will be observable from Earth until about 2040.

Following completion of the Cassini Mission at Saturn, there will be a body of ring and related data that will require decades to digest. As these data lead to conclusions, undoubtedly those conclusions will spawn countless theoretical and observational campaigns and perhaps even follow-on space missions. B. A. Smith [1] attempted in 1984 to predict the types of Earth-based observations that would contribute to improved knowledge of the known planetary ring systems (Neptune's ring system was still a matter of speculation). He accurately predicted that the Hubble Space Telescope would be an invaluable tool for such studies. His statement that "The principal problem with groundbased observations of . . . rings is the limited resolution imposed by the terrestrial atmosphere . . ." did not foresee the development of adaptive optics [2]. Such optics now routinely achieve imaging resolutions comparable with and sometimes better than that of the orbiting Hubble Space Telescope. Learning from his experience, we will not attempt to outline in detail what can and cannot be observed about Saturn's rings from ground-based telescopes, nor what theoretical studies will be pursued, but it is certain that there will be many such observations and studies over the coming decades.

Stellar occultations viewed from Earth continue to be the primary means of studying the narrow rings of Uranus. Stellar occultations also provide useful data on the rings of Saturn and on the arc regions of the Adams ring of Neptune.

When the rings of Uranus were discovered in 1977, and when they were imaged by Voyager 2 in 1986, they were essentially concentric rings surrounding the planet. However, as Uranus continues in its orbit around the Sun, its ring system, like that of Saturn, is edge-on to the Sun twice each 84-year orbit. The next Uranus "equinox" occurs on December 7, 2007. Associated with that equinox, the Earth will cross the Uranus ring plane three times within a 10-month period, on May 2 and August 16, 2007, and on February 20, 2008 [3]. These times will be very useful for observations of the tenuous dusty rings of Uranus, including the two recently discovered rings 2003U1R and 2003U2R.

## 12.3 EARTH-ORBITING TELESCOPIC OBSERVATIONS

The Hubble Space Telescope (Figure 12.1, [4]) continues to operate as of this writing and will continue to do so for several years. Its primary mirror is 2.4 m in diameter. It can resolve details as small as 0.05 arcseconds across. That corresponds to sizes near

**Figure 12.1.** An image of the Hubble Space Telescope (HST) shortly after its release from the bay of the Space Shuttle. HST has undergone several servicing missions since its launch, including one to correct a flaw in the primary mirror shape. HST's contribution to ring studies at all four giant planets is considerable.

**Figure 12.2.** The Spitzer Space Telescope was placed into solar orbit by an unmanned spacecraft. This is an artist's concept of the Spitzer Telescope in orbit. Although few planetary ring studies have yet been carried out using the Spitzer Telescope, it has the capability to collect infrared images and spectra on planetary rings.

150, 310, 660, and 1,050 km at Jupiter, Saturn, Uranus, and Neptune, respectively. While much poorer than achieved by spacecraft encountering or orbiting one of the giant planets, these resolutions are sufficient to reveal changes in ring appearance over the decades since the Voyager encounters. Imagery of the Neptune ring arcs at comparable resolutions recently revealed changes in the relative brightness and positions of several of the arcs [5].

The Hubble Space Telescope orbits Earth at an altitude of 600 km. A newer infrared telescope, the Spitzer Space Telescope (Figure 12.2 [6]), is actually in orbit around the Sun rather than around the Earth. It follows the Earth in its orbit, at a distance that increases about 15,000,000 km each year. While its main goals are associated with astronomy beyond the solar system, it also studies the giant planets circling our own star, the Sun, and could provide temperatures, images, and other data on planetary ring systems.

The Hubble Space Telescope will eventually be replaced by the infrared-optimized James Webb Space Telescope (Figure 12.3, [7]). With a primary mirror 6.5 m in diameter, its light-gathering power and resolution will improve substantially on that

**Figure 12.3.** A mock-up of the James Webb Space Telescope (JWST), scheduled for launch no earlier than 2013, is on display at NASA's Goddard Space Flight Center in Greenbelt, Maryland. JWST is a more capable version of the Hubble Space Telescope and will orbit the Sun in Earth's L2 Lagrangian point.

of the Hubble Space Telescope. However, it is not scheduled for launch until at least 2013. The James Webb Space Telescope will be inserted into the Earth's L2 Lagrangian point [8], nearly 1,500,000 km from Earth (about four times the distance of the Moon), where it will orbit the Sun with the same orbital period as Earth, thanks to the combined gravitational forces of the Sun and the Earth.

## 12.4 SPACECRAFT MISSIONS TO THE GIANT PLANETS

Only two outer planet missions are presently active and in a mission operations mode: Cassini and New Horizons. The Cassini orbital tour of Saturn, its preliminary ring-related discoveries, and its planned future observations were discussed in detail in Chapter 10. The New Horizons spacecraft was launched on January 19, 2006, and passes close to Jupiter on February 28, 2007 [9]. Specific plans for Jupiter observations will likely include a measure of the dust environment near Jupiter, along with five other ring-related studies: (a) ring phase curve observations; (b) high-resolution ring imaging; (c) search for new inner satellites; (d) ring-plane crossing observations; and

(e) studies of quadrant asymmetry (like those seen in Saturn's A ring). The primary goal of the New Horizons mission is to study Pluto and its moons and to study from close range at least one Kuiper Belt object beyond Pluto.

One other outer planets mission is presently in preparation: the Juno Mission will be a Jupiter Polar Orbiter, and is presently scheduled for a launch in 2011. Several others are being considered by the National Aeronautics and Space Administration [10]. Three additional Jupiter missions are currently under study. These include a Europa Geophysical Explorer, with a possible launch no earlier than 2015; a Jupiter Flyby with Probes, with a launch around 2020; and a Europa Astrobiology Orbiter, with a launch about 2035 or later. Of these four Jupiter missions, Juno (the Jupiter Polar Orbiter) offers the greatest opportunity for good ring system data.

Following on the heels of the Cassini/Huygens Mission, a possible Titan Explorer with Orbiter is being considered. The launch date would be no earlier than 2025, and a Titan orbiter spacecraft is not well positioned for Saturn ring studies because of its distance from Saturn and its likely confinement near the equatorial plane of Saturn.

No missions are presently under study for Uranus. A Neptune Triton Orbiter, with unspecified launch date, is presently being considered.

## 12.5 NOTES AND REFERENCES

[1] Smith, B. A., 1984, "Future observations of planetary rings from groundbased observatories and Earth-orbiting satellites", in *Planetary Rings*, edited by Greenberg, R. and Brahic, A., pp. 704–712.

[2] Adaptive optics utilizes an image of a star or other point source close to the direction of the desired target in the sky to make nearly instantaneous, computer-driven changes in the shape of the primary mirror in a telescope, thereby removing the effects of atmospheric turbulence. If the primary telescope mirror in an adaptive optics system is larger than the 2.4-m diameter primary mirror of the Hubble Space Telescope (HST), the images thus obtained can often exceed the sharpness of HST images. The Infrared Telescope Facility and the Keck Telescope atop Mauna Kea on the island of Hawaii both have larger primary mirrors than HST and are equipped with adaptive optics systems.

[3] The Uranus ring crossing dates were determined using the Uranus ephemeris tool available at *http://pds-rings.seti.org/tools/ephem2_ura.html*

[4] The Hubble Space Telescope website is *http://hubble.nasa.gov/index.php*

[5] De Pater, I., Gibbard, S. G., Chiang, E., Hammel, H. B., Macintosh, B., Marchis, F., Martin, S. C., Roe, H. G., Showalter, M., 2005, "The dynamic Neptunian ring arcs: Evidence for a gradual disappearance of Liberté and resonant jump of Courage", *Icarus* **174**, 263–272.

[6] The Spitzer Space Telescope website is *http://www.spitzer.caltech.edu/*

[7] The James Webb Space Telescope website is *http://www.jwst.nasa.gov/*

[8] Lagrangian points are named in honor of Italian–French mathematician Joseph-Louis Lagrange (1736–1813), who first pointed out that an object situated on a line between Earth and Sun at a distance from Earth of about 1% of the distance to the Sun, will remain stably between Earth and Sun indefinitely. That point is called the L1 point. He also defined an L2 point, also along the Sun–Earth line, but outside Earth's orbit at a comparable distance from Earth with the L1 point. The L3 point is at Earth's distance from the Sun, but

precisely on the opposite side of the Sun from Earth. L4 and L5 points, respectively, lead and trail Earth in their orbit by 60°, or 1/6 of Earth's orbital circumference. Other planets also have similar Lagrangian points. The Trojan asteroids orbit near the L4 and L5 points of Jupiter's orbit. Because small bodies (or spacecraft) placed near any of these Lagrangian points remain close to, but wander about, the points, Lagrangian points are sometimes called *libration* points.

[9] *http://pluto.jhuapl.edu/science/jupiterScience/ringsThings.html*

[10] See *http://solarsystem.nasa.gov/missions/index.cfm*

## 12.6 BIBLIOGRAPHY

Greenberg, R., Brahic, A., 1984, *Planetary Rings*, University of Arizona Press; especially chapters by Stone, E. C., "Future studies of planetary rings by space probes", pp. 687–703, and by Smith, B. A., "Future observations of planetary rings from groundbased observatories and Earth-orbiting satellites", pp. 704–712.

## 12.7 PICTURES AND DIAGRAMS

Figure 12.1 *http://hubblesite.org/gallery/spacecraft/06/lg_web.jpg*
Figure 12.2 *http://ipac.jpl.nasa.gov/media_images/earth_trailing2.jpg*
Figure 12.3 *http://www.jwst.nasa.gov/images/fullscale.jpg*

# Glossary

**Albedo** Reflectivity of a body. *Bolometric Bond albedo*: ratio of total flux reflected in all directions and all wavelengths from a planetary body to total incident flux. *Geometric albedo*: ratio of observed planetary brightness at a given phase angle to the brightness of a perfectly diffusing disk with the same position and angular size.

**Ansae** Portion of rings that appears farthest from the disk of a planet. In early Saturn drawings, the "handles" of the rings were so called because they resembled the handles of a vase.

**Apsidal precession** A slow revolving (precession) of the points (apsides) at each end of the major axis of an elliptical orbit around the central planet.

**Back-scattering** Reflecting or scattering radiation (light or radio waves) back towards its source. Reflective particles much larger than the wavelength of the incident radiation are usually back-scattering.

**Bending waves** A moving pattern of wave motion in which particles oscillate normal to the ring plane as a result of the gravitational influence of a nearby satellite in an inclined orbit. The wave is set up at a radial distance in the rings where ring particles orbit in a time period that is a simple integer fraction of the satellite's orbital period; it propagates (toward the planet) due to collective gravitational effect of particles on neighboring regions. The wave pattern turns at the same rate as the orbital motion of the satellite.

**Co-orbital satellites** Satellites which either share the same orbit or which occupy immediately adjacent orbits that change periodically as the satellites approach one another.

**Density waves** Wave motion in a flat disk of particles, generally driven by resonant perturbations by a distant satellite. The wave propagates outward from the planet due to collective gravitational effect of particles (i.e., local surface density variations) on particles in neighboring regions; the wave pattern revolves around the planet at the orbital rate of the perturbing satellite. The theory was originally developed to explain spiral structure of galaxies. Also known as *spiral density waves.*

**Differential rotation** The cause of *Keplerian shear* within extended rings. The laws of orbital motion dictate that ring particles closer to the planet move at greater speeds and have shorter orbital periods than their slightly more distant neighbors. *Differential rotation* might more appropriately be called *differential revolution.*

**Dynamic oblateness** A measure of the extent to which mass has been shifted from the polar regions of a (spinning) body towards its equator.

**Eccentric** Non-circular, elliptical. Orbital eccentricity ranges from 0 (circular), to 0.999 (very elongated ellipse). A parabola has an eccentricity of 1.0; an eccentricity greater than 1.0 is characteristic of a hyperbola.

**Equivalent depth** A measure of the number of particles passing a given point in the ring per unit time. It is obtained by multiplying the physical width of the ring by its average optical depth. For the variable-width eccentric rings of Uranus, equivalent depth remains almost constant around a given ring.

**Equivalent width** The width-integrated optical depth of a ring. For rings with very small optical depths, the equivalent width is very nearly equal to the equivalent depth.

**Forward-scattering** Preferentially reflecting or scattering radiation (light or radio waves) in a direction away from its source. Particles with sizes comparable with the wavelength of the incident radiation tend to forward-scatter that radiation. The degree of forward-scattering is therefore a diagnostic tool in particle-size determination.

**Gravity wakes** Transient streamers which form when clumps of particles begin to collapse under their own *self-gravity* but are sheared out by differential rotation. This phenomenon is believed to be the source of azimuthal asymmetry in Saturn's A ring.

**Gravitational resonance** Same as *Lindblad resonance.* A repeated gravitational force on a ring particle in the same direction and at the same point in its orbit. This force occurs on ring particles that are at a radial distance such that their orbit period is a simple fraction of that of a nearby satellite.

**Keplerian shear** Shearing motion of an ensemble of particles, each on a nearly circular, Keplerian orbit. Orbital velocity decreases as orbital radius increases, yielding shear. *Viscous drag* on such shear, due to ring-particle collisions, plays a key role in ring processes.

**Kirkwood gaps** Gaps in the semimajor-axis distribution of main-belt asteroids, located at solar distances where their orbital periods are simple integer fractions of Jupiter's orbital period.

**Lagrangian points** Five gravitationally stable points (relative to a planet orbiting the Sun) where a small third body can orbit the Sun at the same rate as the planet. *L1* is between the planet and the Sun, *L2* is along the Sun–planet line, but exterior to the planet, and *L3* is at the same planet–Sun distance, but on the opposite side of the Sun. The *L4* and *L5* points, respectively, lead and trail the planet by 60° in its orbit.

**Laplace plane** The near-equatorial (ring) plane as altered by the gravitational effects of the Sun and large satellites. Close to a giant planet, the *local Laplace plane* is essentially the planet's equatorial plane. Distant satellite orbital inclinations are generally referred to the *local Laplace plane* rather than to the planet's equatorial plane.

**Lindblad resonance** (= *gravitational resonance*) A repeated gravitational force on a ring particle in the same direction and at the same point in its orbit. This force occurs on ring particles that are at a radial distance such that their orbit period is a simple fraction of that of a nearby satellite. *Inner Lindblad resonances* occur when the satellite is exterior to the given ring. *Outer Lindblad resonances* occur when the ring is exterior to the satellite's orbit.

**Lorentz resonance** A repeated electromagnetic force on an electrically charged ring particle, nudging the particle in the same direction and at the same point in its orbit. *Lorentz resonances* are especially important for tiny ring particles whose charge-to-mass ratio is high and whose orbit periods are a simple integer fraction of the rotational period of the planet's magnetic field.

**Mass density** The mass per unit area of the ring material, integrated through the thickness of the ring. Sometimes called *surface density*.

**Moonlet wakes** Local disturbances in the ring structure caused by the gravitational influence of embedded satellites. If the satellite (moonlet) is large enough to clear a gap in the rings, the moonlet wakes become *edge waves* that precede the satellite on the inner edge and trail the satellite on the outer edge. For smaller satellites, the "gap-less" wakes have been nicknamed *propellors*.

**Nodal regression** For a ring inclined to the planet's equator, the points at which the ring crosses the equator (nodes) slowly move around the planet (regress) in a direction opposite to that of the ring's orbital motion.

**Occultation** The passage of one celestial body in front of another. For stellar ring occultations, the ring passes in front of a star as seen by the observer; for radio ring occultations, the radio transmitter passes behind the rings as viewed by radio antennas on Earth.

**Opposition effect** Abrupt increase in brightness of a body or ring near zero phase angle. Also known as the *zero phase effect* or *opposition surge.*

**Optical depth** The ratio of the intensity of radiation (light or radio waves) incident on a ring to that emerging from the opposite face of the ring, expressed as a natural logarithm. If the reduction in intensity is by a factors of $e$ (= 2.718), the ring is said to have an optical depth of 1. *Normal (or normalized) optical depth* is optical depth corrected for oblique (non-vertical) viewing. In ring studies, the terms *optical depth* and *optical thickness* are generally used interchangeably. These terms generally refer to a particular wavelength.

**Optical oblateness** (polar flattening) The ratio of the difference between equatorial and polar diameters to the equatorial diameter. A sphere has an oblateness of 0; an infinitely thin disk has an oblateness of 1.

**Phase function** The variation in brightness of a target as the *phase angle* (the angle between Sun and observer as seen from the target) varies between 0° and 180°. The directional distribution of reflected (or scattered) radiation. The *phase angle* is the supplement of the *scattering angle* (the angle between the incident ray and the emerging ray); in other words, the sum of the phase angle and the scattering angle is always 180°.

**Photoelectric effect** Ejection of electrons from a substance by the energy contained in incident electromagnetic radiation, especially by sunlight.

**Poynting–Robertson drag** A loss of orbital angular momentum by tiny ring particles associated with their absorption and reemission of solar radiation. Also known as the *Poynting–Robertson effect.*

**Pulsational instability** A term used to describe irregularly spaced, fine-scale structure in optically thick rings. The process relies on a combination of *viscosity* and *self-gravity* of ring material to produce this fine structure. Also known as *overstability.*

**Radial spokes** Short-lived (generally lasting less than 24 hours) radial features that periodically appear over the outer half of Saturn's B ring, when the ring tilt angle is small. These features revolve at the same rate as the planet's magnetic field and maintain their shape over much of the course of their existence even though they extend tens of thousands of kilometers across the rings. It is believed that the tiny particles that make up these spokes are electrically charged and temporarily "frozen" into the planet's magnetic field.

**Resonance** Selective response of any periodic system to an external stimulus of the same frequency as the natural frequency of the system. Ring particles can have a resonant response to the periodic driving force of a satellite's gravity.

**Roche limit** The minimum distance at which a zero-strength satellite, influenced by its own gravitation and by that of the planet it orbits, can exist. Such a satellite, with the same mean density as its planet but much smaller in size, will break up at 2.44 times the radius of the planet.

**Shepherding satellite** A satellite in orbit near the edge of a planetary ring, that exerts a gravitational torque and thus maintains the integrity of the planetary ring by preventing it from spreading.

**Thermal inertia** The tendency of a body to resist a change in temperature. A body with a low thermal inertia requires very few calories to change its surface temperature. A low thermal inertia material tends to be thermally insulating, so that the surface temperature changes readily, but those changes are not conducted to depth within the material.

**Tidal forces** The mutual gravitational forces between a small body and the larger body it orbits. The Moon's tidal forces produce high and low tides on the Earth; in turn, the Earth's tidal forces deform the Moon, elongating its diameter along the Earth–Moon direction.

**Viscosity** The resistance that a fluid system offers to flow when it is subjected to a shear stress. For a dense ring (like Saturn's A or B ring), interparticle collisions create effects that mimic *viscosity* in fluids, reducing relative velocities, damping out spiral wave motion, and resisting non-zero inclinations in ring-particle orbits. See *Keplerian shear*.

# Index

(*italics* refer to images or tables; **bold** refers to detailed discussions)

Printing: Mercedes-Druck, Berlin
Binding: Stein + Lehmann, Berlin